WEATHER RADAR HANDBOOK

TIM VASQUEZ

First edition
January 2013

Color revision
November 2015

Copyright ©2013 Weather Graphics
All rights reserved

For information about permission to reproduce selections from this book, write to Weather Graphics Technologies, P.O. Box 450211, Garland TX 75045, or to support@weathergraphics.com . No part of this publication may be reproduced, stored in a retrieval system, or transmitted by any means without the express written permission of the publisher.

ISBN 978-0-9969423-1-7

Printed in the United States of America

Weather Graphics Technologies
P.O. Box 450211 Garland, TX 75045
(800) 840-6280 fax (206) 279-3282
Web site: www.weathergraphics.com

servicedesk@weathergraphics.com

CONTENTS

1 History
1.1. A prelude to weather radar / 1
1.2. Weather radar in the United States / 2
1.3. Weather radar around the world / 9

2 Fundamentals
2.1. Properties of electromagnetic waves / 15
2.2. Radar energy / 18
2.3. Radar energy / 22
2.4. The radar beam / 26
2.5. Volumetric scanning / 32
2.6. Visualization / 34
2.7. WSR-88D / 39
2.8. TDWR / 42

3 Reflectivity
3.1. Problems affecting reflectivity data / 47
3.2. Precipitation / 50
3.3. Thunderstorms / 52
3.4. Squall lines / 59

4 Velocity
4.1. Velocity processing / 63
4.2. Display standards / 69
4.3. Volumetric signatures / 70
4.4. Localized signatures / 71
4.5. Spectrum width / 77
4.6. Wind profile products / 79

5 Polarimetry
5.1. The basics of polarimetry / 87
5.2. Differential reflectivity (Z_{DR}) / 88
5.3. Correlation coefficient (CC, rho, ρ_{HV}) / 90
5.4. Differential propagation phase (ϕ_{dp}) / 91
5.5. Specific differential phase (K_{dp}, KDP) / 91
5.6. Linear depolarization ratio (LDR) / 92
5.7. Polarimetric algorithms and derived products / 92
5.8. Differential diagnosis / 95
5.9. Features and phenomena / 97
5.10. Filtering / 99

6 Derived Products
6.1. Composite reflectivity / 103
6.2. Vertically integrated liquid / 104
6.3. Echo tops / 105
6.4. Precipitation estimation / 106
6.5. Storm detection algorithm / 109
6.6. Mesocyclone detection algorithms / 110
6.7. Tornado detection algorithms / 111
6.8. Hail detection algorithms / 111
6.9. Storm relative velocity / 113
6.10. Velocity azimuth display / 113
6.11. VAD wind profile / 113

7 Forecast Integration
7.1. Quiescent weather / 117
7.2. Boundaries / 118
7.3. Stratiform precipitation / 119
7.4. Convective precipitation / 119

Appendix
Common radar abbreviations and variables / 135
Range-height diagram / 136
Storm motion nomogram / 137
NEXRAD product codes / 138
Summary of typical polarimetric parameters / 139
United States NEXRAD coverage map / 140
List of WSR-88D builds / 142
WSR-88D Volume Coverage Pattern (VCP) reference / 144
WSR-88D Volume Coverage Pattern (VCP) / 144
References & recommended reading / 147

Index / 149

Introduction

The year 2013 will be a great one for meteorology, because in one large swoop it brings the addition of polarimetric products to the national radar network. Every person involved in any significant capacity in meteorology, from the dedicated hobbyist right up to the science and operations officer at a National Weather Service office, finds theirselves not only with powerful new tools but also a challenging learning curve.

This network upgrade has been the catalyst for me to write my first dedicated book on radar meteorology. It also comes on the 20th anniversary of my own immersion in WSR-88D equipment as our station phased out an FPS-77 radar and brought in a NEXRAD PUP and UCP. Since then the systems have taken on improvements that I never would have thought possible. One thing is certain: there is always more to learn about, and it's never quite enough. In this book I have tried to distill knowledge and techniques applicable to operational forecasting without allowing the core material to grind down to a halt in complicated topics. My goal is to make this title as readable and *useful* as possible.

The contents of this book will frequently refer to the United States radar network. Though this dovetails with the fact that most of my readership is American, the unfortunate fact remains that for whatever reason, the United States remains the only country in the world that offers unrestricted public access to all of its meteorological data; not only radar but models and satellite data and all the binary source data that makes it up. American students, educators, pilots, small businesses, and hobbyists all have equal footing with professionals, industry, and institutions in exploring all of the Doppler and dual-pol radar information that exists and finding uses for it all. If there is a shift in data access policies by any national governments, I will gladly include those radar networks in future editions of this book. Either way, the underlying principles of radar analysis and interpretation, as well as the elements of weather forecasting, are universal. All information in this book will be useful regardless of where the reader lives.

I have avoided delving into too much technical information on the WSR-88D, since this would fill the pages with a lot of detailed data that is not essential to the material. Readers may refer to FMH-11, listed in the references, as an authoritative source for details on this radar system.

As with all of my projects, this is an ongoing work, and any corrections or suggestions are gladly appreciated for future editions. Any errors we find will be posted at <www.weathergraphics.com/wrh/>.

Tim Vasquez
January 15, 2013
Norman, Oklahoma

In every branch of knowledge the progress is proportional to the amount of facts on which to build, and therefore to the facility of obtaining data.

-- James Clerk Maxwell, 1851
father of electromagnetism

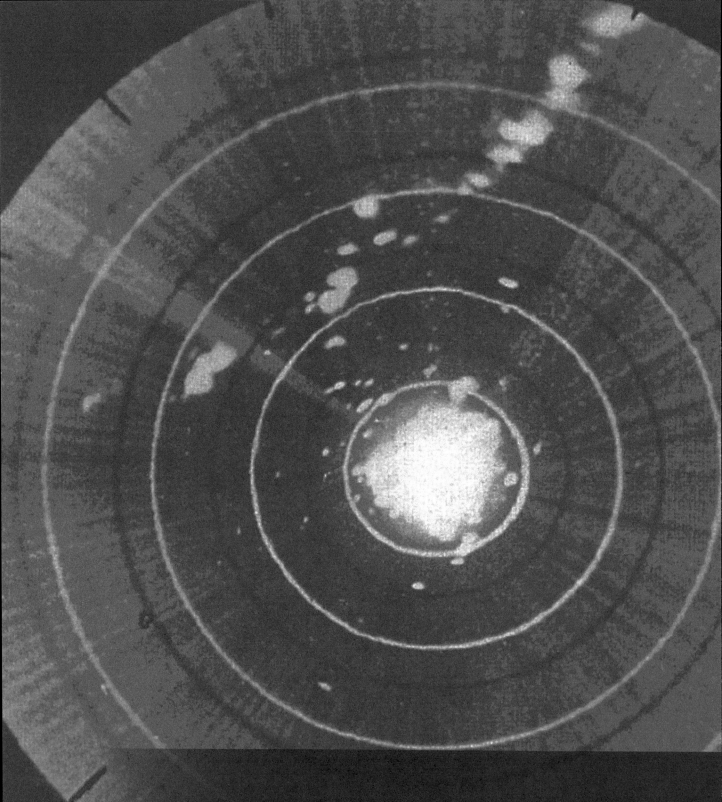

1 HISTORY

1 | HISTORY

Storm detection in its current form was not even conceivable during the opening years of the 20th century. At the time there were experimental storm detection networks in Germany that used visual sightings and telegraph reports. Fast forward to the 21st century. The ability to sample characteristics like droplet sizes, water loading, particle types, and movement at any given point in the atmosphere almost certainly ranks among one of the great breakthroughs of the modern era. Many of the greatest advances in storm detection have occurred only in the past 25 years!

Radar, which stands for *radio detection and ranging*, is built on the foundation of radio electronics, which in turn is deeply rooted in the science of electromagnetism, one of the most fundamental forces of the universe. Conventional severe weather forecasting deals with understanding heat exchange processes in a target area and sifting through a plethora of tools to understand how they will translate to thunderstorm activity. However, radar technology requires different skill sets: understanding radio waves, electromagnetism, electronics, and the engineering of radar equipment itself. This book will focus mostly on the parts that are important to operational forecasting.

1.1. A prelude to weather radar

A popular story that still lingers in American history textbooks recounts Benjamin Franklin as a discoverer of electricity in an apocryphal 18th century kite and thunderstorm experiment. However knowledge of electromagnetism actually goes back thousands of years. Around the 6th century BC the Greeks had successfully built a static electricity machine, in which two wheels ground against each other in opposite directions; one wheel lined with cat fur and the other with amber. This apparatus was able to create large sparks. Other curiosities such as lodestones, catfish, electric eels, and of course lightning were all known to the ancient world. As the bedrock of chemistry and atomic theory had not yet been mapped out, the magic behind these effects were attributed to divine intervention and supernatural life within the objects.

The connection between electricity and magnetism and a deeper understanding did not emerge until after the Middle Ages. By the early 19th century, the properties of insulation, resistance, and capacitance were well understood. Metallic wire was capable of transmitting a charge representing information, and short-range telegraphs were in service by 1833 with national networks spreading out by the 1850s. In the United States, the telegraph linked the East Coast and West Coast in 1861, putting the short-lived Pony Express out of operation. By 1866, a reliable transatlantic telegraph cable was in continuous service. All of this infrastructure laid the groundwork for synoptic weather forecasting.

Radio communication, known initially as "wireless", would not be known for decades. Many ancient Greek philosophers and later-day

Title image
WWII-era radar image at Lakehurst, New Jersey on 16 July 1944, at 1842 UTC. It shows an area of showers developing along a cold front. *(U.S. Navy)*

The power of radar

The WSR-88D transmits at a power of 750 kW, comparable to the super-power AM radio stations that can be heard almost nationwide, though on a different band. All this energy is focused into a beam less than 2 degrees in width.

researchers such as Galileo had long argued about the nature of light, making various attempts to measure its speed to quantify its behavior. Danish astronomer Rømer obtained a fairly accurate value in 1676 by observing the clockwork behavior of Jupiter's moons from different positions in Earth's orbit. But a major milestone came in 1865 with Scottish physicist James Clerk Maxwell's proposal that electricity and magnetism were intertwined, and that because the speed of electromagnetism was of the same as light, that light was just a form of electromagnetism.

When telegraphy was first developed, it was found that strong voltage in a conductor could produce a weak resonant charge in conductors in the same room: an effect called induction. The effect was weak, and for decades it was thought to be impossible to use this weak effect to transmit information. In 1887, German physicst Heinrich Hertz successfully used Maxwell's theories to develop an *oscillator* which would alternate the voltage to radiate electricity, rather than radiating it as a spark or as an inductance response to high voltage. The apparatus also included a dipole antenna. Unfortunately Hertz's quickly health declined and he died several years later at the age of 36.

Both Serbian-American scientist Nikola Tesla and Italian inventor Guglielmo Marconi recognized the ability of "Hertzian waves" to carry information and perhaps be received over long distances. Both Tesla and Marconi developed devices were built in the last few years of the 19th century that could transmit telegraphic information over great distances, and by 1900 the first voice transmission was made. The first decade of the 20th century saw an explosion in wireless telegraphy technology, and by the end of the decade transceivers were installed on most large ships. Audio broadcasting would not become commonplace until the 1920s.

Hertz's experiments in the 1880s clearly demonstrated the ability of substances to reflect electromagnetic energy. The potential for radio waves to "find" objects was known as far back as 1897 and various systems were considered to facilitate ship safety, but no real progress was made. With aircraft filling the skies in the 1930s and taking on tactical and strategic significance, military researchers in the United States, Germany, and Russia developed the framework for radar technology and developed the first generation of experimental sets. This technology was rapidly fielded with the outbreak of World War II in 1939 and rapidly matured. Weather echoes became a frequent annoyance for technicians and commanders during combat operations, but of course there were also those who looked at the weather echoes and saw the future.

1.2. Weather radar in the United States

The extensive experience with radar at the research and operational level during WWII led to a myriad of forks in the technology that tailored radars to highly specific applications such

Figure 1-1. James Clerk Maxwell (below) is credited for his 1865 theory of the electromagnetic field. This paved the way for the discovery of radio, and later, radar. *(Master and Fellows of Trinity College, University of Cambridge).*

as air traffic control, airborne navigation, military defense, and of course meteorology. In meteorology itself, there have been at least four significant revolutions affecting the operational forecaster. The first is the fielding of radars solely for forecasting use, which was in itself revolutionary. The second involves the 1961 deployment of highly specialized meteorogical radars designed for network use. The third generation introduces the use of velocity data in solving forecast problems. The fourth generation, which was unfolding at the time of this writing, adds polarimetric diagnosis to the previous three advances.

1.2.1. First Generation Weather Radar. During WWII, radar technology was secret and expensive, which limited research and experience. However under the auspices of the War Department a handful of institutions, most notably the Massachussetts Institute of Technology (MIT), were exploring the effects of weather on radar, not only helping American forces mitigate its impact during combat operations but also developing a solid body of knowledge on weather's radar characteristics. Parallel work in radar meteorology was taking place amongst all superpowers, but especially in Great Britain, where in later years radar first identified the supercell.

One of the first radar forecasting initiatives can be traced back to 1945 at the Indian bases serving as the western terminus of "The Hump", an aerial supply line from east India into China to repel Japanese forces. Several APQ-13 radars were routinely sent up on bombers to watch the northwest horizons for squall lines, which were a major hazard. An idea was floated to install the radars on the ground. This proved to be a success, and by mid-1945 a ground-based radar network was in place across the Assam region. This provided military forecasters with unprecedented information on the location of storms in the western portions of the Hump.

World War II's closure in August 1945 allowed for the declassification of radar technology and unhindered growth of the science of radar meteorology. The Weather Bureau acquired 25 ex-Navy radar units and put them to use in 1946. Programs like the Thunderstorm Project revealed many of the inner workings of thunderstorms.

The first radar specifically designed for meteorological use was the CPS-9 X-band radar. However the CPS-9 was developed under a military contract, and the units were installed in the mid-1950s only at military bases. Though they greatly improved warning capability for the Department of Defense and provided them with a second-

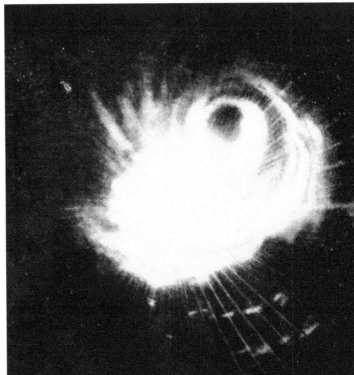

Figure 1-2. The battlefields of World War II comprised one of the birthplaces of radar meteorology. Shown here is Typhoon Cobra east of the Philippines on 18 December 1944, viewed by a U.S. Navy ship caught in the maelstrom. The typhoon, believed to be a Category 4 storm at the time, was responsible for 790 deaths as it passed over Task Force 38, sinking three destroyers, damaging nine other ships, and ruining hundreds of planes. The force had been headed for the Philippines to oust Japanese air forces from Luzon. *(U.S. Navy)*

VIP Levels

During the 1970s through the mid-1990s, radar data from the second generation radars was expressed as a "VIP Level". Shown here is a conversion chart. Although VIP 1 is 0 to 29 dBZ, the WSR-57 was not sensitive enough to detect less than 18 dBZ.

VIP	dBz	Typical color
1	0-29	Light green
2	30-37	Dark green
3	38-43	Yellow
4	44-50	Beige
5	50-56	Red
6	57+	White or magenta

generation capability, there was no widespread availability of the data to civilian forecasters and researchers.

Throughout the 1950s, the Weather Bureau continued to use its WWII radar units. While the U.S. military had successfully issued tornado warnings and had a state-of-the-art weather radar network, the Weather Bureau continued to position itself as strictly a provider of daily forecasts and a steward of agricultural and industrial interests. Although it was well aware of severe weather forecasting techniques that began emerging in the 1920s, it saw local storms as outside its scope of responsibility and only by 1943 did it agree to relay warnings if storms were already underway.

1.2.2. SECOND GENERATION WEATHER RADAR. A combination of bad press and congressional pressure forced the Weather Bureau in 1952 to accept the burden of providing warning services. The first innovation was the Severe Local Storms Unit (later SELS, NSSFC, then SPC) in Kansas City. Though some of the agency's primitive WWII radars were upgraded, it was a series of damaging hurricanes in 1955 which led Congress to set aside funding for the development of a next-generation radar network. This radar, designed by Raytheon, became known as the WSR-57. It was deployed across most areas east of the Rockies starting in 1961. This radar would form the backbone of most operational radar forecasting in the United States for the next 35 years.

The WSR-57, while revolutionary, did not resemble today's automated digital radars. A radar operator was physically assigned to each radar site in a full-time capacity. The operator made manual and special observations of radar images, transmitting general descriptions of the echoes in a coded form on teletype. This code form was known as RAREP. In Washington DC, analysts assembled all of the RAREPs every hour into a national radar composite, which was transmitted over Weather Bureau fax circuits starting around 1961. They greatly helped with flight safety and general forecasting, and even after 50 years the charts are still produced very similar to their original form!

Interrogation of storm intensity was a continuing problem. Originally, radar data was displayed on crude phosphor screens that gave very poor differentiation between weak and strong echoes. The first efforts to depict intensity used electronic techniques to create a light-dark separation or filter the weak echoes out, creating a sort of contour effect on the radar screen. This system, known as the Video Integrator and Processor (VIP), was added to network radars in 1968. By the 1970s, the advent of computer processors and display technology allowed radar data to be displayed in a conventional colorized format on CRT displays, known as DVIP (digital VIP).

By the 1970s, RAREP was proving to be insufficient for operational needs, especially with the advent of numerical forecasting. The RAREP code only gave general depictions of where echoes were and where the strongest ones were located, and being a manually assembled product it was subject to human error. A scheme known as MDR (Manually Digitized Radar) was fielded in November

1973 in which the radar operator manually coded echoes onto a 40 km grid using an intensity of 0 to 6. This gave a sort of "poor-man's" digital radar display but allowed for a substantial upgrade in the hourly national radar composite chart, since it could be automatically generated by specialized FORTRAN programs. Even as of 2012 these composites are still available in much the same form as they were over 30 years ago.

Storm-scale meteorology was making huge strides during the 1960s, but the transmission of detailed imagery that could allow diagnosis of individual storms was simply not available to anyone except the radar operator at the site. This separation between forecasters and radar sites prompted the Weather Bureau to develop the WBRR (Weather Bureau Radar Remote) system, which photographically slow-scanned the radar display, including the technician's markings, and transmitted it over standard dial-up or dedicated telephone lines for remote display on television displays or film recorders. This system was widely deployed by 1968. In the late 1970s, digital data networks and modem technology began allowing for digital processing of radar data, and this allowed NWS forecast offices to get pure, high-quality radar imagery at the forecast desk. By 1982, when the Weather Channel appeared on cable networks, the DVIP displays, mostly provided by Kavouras, were a mainstay of weather programming and would be seen by millions of Americans regularly throughout the 1980s. Meanwhile many users, including air traffic controllers, other forecast offices, industrial meteorologists, television stations, and corporate users enjoyed the ability to

Figure 1-3. Example of a second-generation radar: a WSR-74C radar unit. This example, photographed in April 1989, was owned by an individual living in a rural home east of Dallas. The equipment was bought as surplus and licensing was obtained. This shows the PPI scope, which doubled as an RHI scope for cross sections. The radar dish was controlled by hand cranks on the console on the right. The system contained a DVIP display which produced an electronic image seen on the television set at the center of the picture. One can only wonder whether someday in the future a hobbyist will own a decommissioned WSR-88D! *(Tim Vasquez)*

Figure 1-4. The NSSL 10 cm Doppler radar at Norman, Oklahoma, as seen on 26 April 1999. This radar was installed in 1969, having been a surplus FPS-18 radar from the Distant Early Warning (DEW) Line near the arctic circle. It became one of the first S-band Doppler radars in the world. It was scrapped a few years after this picture was taken and replaced with the new NSSL multi-function phased array radar (MPAR). It is a fitting end that the first third-generation makes way for the first fifth-generation radar! The KOUN WSR-88D testbed radar can be seen in the background. *(Tim Vasquez)*

"dial into" any radar across the country and see the DVIP feed. This was perhaps the crowning achievement of the WSR-57 network and carried the National Weather Service into the 1990s.

1.2.3. THIRD GENERATION WEATHER RADARS. Though NEXRAD is often thought of as an achievement of the 1990s, its roots go all the way back to WWII when engineers recognized the potential for radars to measure the Doppler shift, a change in the returned frequency or wavelength caused by differences in velocity between the radar site and the target. Some of the very first meteorological tests were done in the late 1950s in a cooperative experiment between the Weather Bureau and Cornell University. This modified radar detected 205 mph winds in a tornado in the central United States. Though this was an exciting prospect, the technological achievements needed to field Doppler radar in an operational setting were not in place.

Research on Doppler weather radars continued through the 1960s and 1970s at the Air Force Cambridge Research Laboratories, the National Severe Storms Laboratory, and the National Center for Atmospheric Research, and this established the fundamentals of vortex signatures, techniques for pulse pair processing, and analysis of volume scans. At the same time, technological advances such as powerful computing technology and high-resolution color displays were brought into the fold of Doppler radar equipment.

Studies conducted by the Environmental Research Laboratory in 1976 concluded that it was not feasible to simply upgrade the existing second-generation radar network to add Doppler capability. In 1977, a federal

Figure 1-5. WSR-88D radome at Kauai (PHKI) as seen on 9 February 2006. Nearly half of all WSR-88D sites use the tallest available version of the antenna tower, including the site shown here. It reaches a height of 30 meters (98 feet) in order to clear the rugged terrain near the site location. *(Nita Patel/NOAA)*

committee made up of the National Weather Service, the Department of Defense (DoD), and the Federal Aviation Administration convened to plan a "next generation radar" known as JDOP (Joint Doppler Operations Project). This formally became known as NEXRAD (Next Generation Radar) in 1979, and by 1981 the first technical specification document was released. The remaining specifications were finalized in 1986, the production name was locked down as WSR-88D in 1988, and Unisys was awarded the production contract. The first prototype radar in Oklahoma City was installed in late 1990.

After a period of extensive testing, deployment of the WSR-88D network took place from 1993 to 1997, and the older WSR-57 and WSR-74 radars at each site were phased out once a nearby WSR-88D radar came online. As of 2012, WSR-88D units were in service at 154 locations in the United States and at 5 DoD sites in Okinawa, South Korea, Guam, and the Azores. An additional 23 radars had been planned for installation by the DoD at bases in Europe, Japan, and the Philippines but were canceled due to budget constraints, and in the latter two, an inability to obtain use of the radio frequency and a base closure, respectively.

An enhancement to third-generation radars called "dual Doppler" has been the subject of research for decades. The Achilles heel of a single-station Doppler radar is that it cannot measure the component of motion perpendicular to the antenna, forcing forecasters and analysis algorithms to make many inferences about what the radar is showing. By combining two Doppler radars, all motions within the combined volume can be solved. Since this type of radar system is not yet used operationally, it is not covered in this book.

1.2.4. FOURTH GENERATION WEATHER RADARS. Though polarimetric ("dual pol") radars are not a complete replacement of existing technology, they will have a profound effect on operational forecasting during the 2010s. In that respect, they are considered to be the fourth-generation weather radar. These radars transmit electromagnetic waves that are polarized in two perpendicular planes, and the returning signals are compared and contrasted to provide inferences about particle composition and distribution within the storm. This can have an enormous impact on the diagnosis of winter weather situations and important processes taking place within thunderstorms. The additional data also improves rain and snowfall estimates and can even give meaningful information on non-meteorological targets such as dust, insects, and birds.

It was recognized as far back as 1945 that materials and targets had the potential to return electromagnetic energy with unique polarimetric signatures. No technology existed, however, that could accurately process and display this kind of radar information for forecasting use. Some early research began during the 1980s which quickly established the potential for polarimetric radar to provide valuable information. It wasn't until the 1990s, however, that computing technology was fast enough to process polarimetric data in real-time.

In 2003, the National Weather Service Office of Science and Technology tasked the National Severe Storms Laboratory to investigate polarimetric radar data and the suitability of upgrading the WSR-88D network, and the success of this program led to the funding of a network upgrade. In March 2011, Vance AFB, Oklahoma became the first polarimetric radar to become fully operational. At press time, nearly all United States WSR-88D sites had been upgraded to polarimetric radars, with the remaining work slated for completion by May 2013.

1.2.5. FIFTH GENERATION WEATHER RADARS. The next major milestone is likely to be the development of multifunction phased array radar (MPAR). These radars eliminate many moving parts, using wave interference techniques rather than motors to "steer" the electromagnetic beam. A volumetric scan can also be completed much more quickly, allowing a faster flow of radar real-time information. Phased array radar is a mature technology and has been in widespread use for military applications for decades.

There is no plan yet in the United States for a replacement of the existing weather radar network. However it is recognized that MPAR radars are the most likely candidate for follow-on technology, and there is the capability to use the same radar unit for meteorological, air traffic control, and air defense purposes, providing a simplified and unified solution for the FAA, DoD, and NWS by the 2030s.

Research into such a system has already begun. A phased array weather radar became operational in September 2003 at the National Severe Storms Laboratory in Norman, Oklahoma. Already the radar has demonstrated its ability to sample storms at a much higher temporal resolution, showing intense, fast-moving processes like microbursts in a minute-by-minute fashion, contrasting sharply with the 6-minute refresh rate of the WSR-88D. Another fifth-generation technology might be considered to be "filler radars", which are relatively cheap units that can fill the gap between network radars and be mounted on cellular phone towers.

The MPAR website
A website for the project and images that are available when the radar is operating can be found at <www.nssl.noaa.gov/tools/radar/mpar/>.

1.3. Weather radar around the world

As stated in the introduction, most of the focus of this book is on the United States. This is not simply a bias due to the author's domicile and the book's readership, but due to the fact that nearly all non-U.S. countries restrict public radar data to images of varying quality, without access to the underlying binary data it is comprised of, and in the great majority of cases only reflectivity data at a single tilt is offered. Most of the radars described here operate in the C-band unless otherwise noted.

1.3.1. NORTH AMERICA. The United States network, consisting of about 150 dual polarization S-band radars, has already been covered thoroughly in the previous section. Canada operates a unified radar network which grew organically out of individual sites at forecast offices and were tied into a national network during the 1990s. These radars were upgraded to Doppler capability starting in 1998. Funding is already in place to add polarimetric capability during the 2010s. There were 31 radars as of 2012, but no coverage exists in any of the northern provinces or in the arctic regions, and radar density is barely adequate in populated regions to provide detailed severe weather analysis. Mexico operates 13 conventional weather radars of various makes and models. Doppler capability was added in the early 1990s. Many of the Mexican radar images are available on the Internet and useful for hurricane forecasting, though content is occasionally unavailable. Radar is also

Figure 1-6. **Experimental phased array weather radar** being installed at the National Severe Storms Laboratory in Norman, Oklahoma. This technology may form the backbone of storm detection and weather forecasting in the 2020s or 2030s. *(NOAA)*

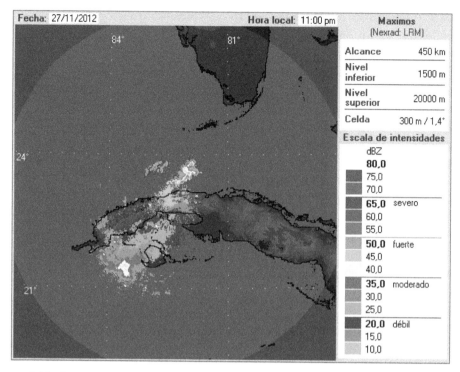

Figure 1-7. **Cuban weather radar**, surprisingly, is on the Internet. Even more surprising is that these consist of Soviet and Japanese radars from the 1970s and 1980s that were modified and upgraded to provide digital processing capability. While they lack Doppler capability and are functionally equivalent to radars used by the National Weather Service in the 1980s, they can be highly useful for monitoring hurricanes. Shown here is the Casablanca radar located just south of Havana, showing dissipating thunderstorms off the coast in the pre-dawn hours of 27 November 2012. *(Institute of Meteorology, Cuba)*

available from some Caribbean sites, most notably Cuba, which operates seven conventional radars. A few isolated radar sites operate elsewhere throughout the Caribbean, such as in Belize, Panama, and Jamaica.

1.3.2. SOUTH AMERICA. Owing to political turbulence and a lack of funding, weather radar networks were slow to develop in Latin America, and where they existed they tended to operate under the auspices of the military. This situation changed during the 1990s with growing economic prosperity and reorganization of many of the weather agencies. The only radar network that exists is in southern Brazil, owing to its high population density. Elsewhere, individual sites exist in cities such as Rio de Janeiro, but overall the continent is too extensive and cost-prohibitive for a network such as that in North America and Europe, and in Chile where a weather radar network is practical, mountainous terrain presents a formidable challenge. Western and northern South America remains completely "dark" in terms of weather radar.

1.3.3. EUROPE. The high population density of Europe and the frequent occurrences of heavy rain and winter weather led many countries to develop national radar networks during the 1950s and 1960s. These were operated independently, but the rapid growth of the European Community during the 1990s led to a dramatic increase in cooperation between weather agencies. A milestone was reached in 1999 with the development of the Operational Program on the Exchange of Weather Radar (OPERA) network, an initiative to network and harmonize the data from all weather radars across Europe. As of 2012 the OPERA network was made up of 224 participating radars, most of them dual-polarization Doppler radars but some of them made up of older conventional units. Although the state of technology in some regions compares closely to that of the United States, public access to the data is widely curtailed in most European countries except as images that show generalized locations of precipitation. The only weather radar network in the former Soviet republics is a six-station network operated by Roshydromet which has covered the Moscow region since the 1980s. The immense size of Russia has been an impediment to its expansion, but Russia is is planning to establish a new network by 2018 consisting of 140 "DMRL-C" dual-polarization Doppler weather radars. This would provide dense coverage in the highly populated western part of the country, with spotty coverage eastward through Siberia at key cities and industrial centers.

1.3.4. ASIA. Because of the high population density of southern and eastern Asia, and in many cases because of the susceptibility to tropical cyclones, radar sites have been operated by many of the weather agencies since the 1950s. Japan and Taiwan were one of the first to develop a modernized weather radar network in the 1980s, followed by South Korea and Thailand. The sudden growth of China's radar network in the 2000s is especially noteworthy: as of press time it consists of 158 Doppler weather radars (mostly C-band in the north and west and S-band elsewhere), with 58 more slated for completion. From an operational standpoint, the system compares favorably to that of the United States WSR-88D network. Indonesia, the Philippines, Bangladesh, and India were all in the process of developing basic radar networks at press time.

1.3.5. AFRICA. Due to unstable political climates and limited government budgets, weather radar networks are largely nonexistent. The exception is South Africa, which fielded 11 conventional radars during the 1980s and later partnered with Botswana and Mozambique to expand the network northward and eastward. Starting in 2010 the country began an aggressive upgrade which is expected to provide forecasters with 18 dual-polarization S-band Doppler

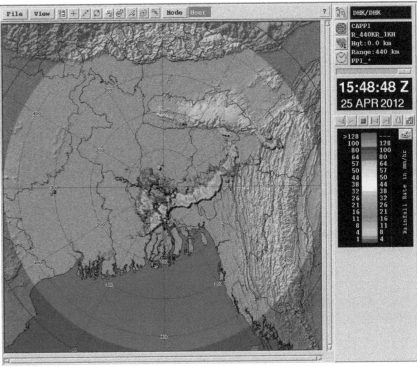

Figure 1-8. Weather radar coverage of India and Bangladesh, is improving. The border area of the two countries is a localized hot spot for supercell development due to the warm Bay of Bengal waters and cool northwesterly upper level flow aloft ducting around the Himalayas. Here a northwest-flow MCS is seem moving southeastward across Kolkata and surrounding areas. (India Meteorological Department)

radar sites. A network of six Doppler weather radars were also reported to be under construction in Nigeria in 2009. The rest of Africa is completely currently devoid of weather radar coverage, with the exception of Morocco and a handful of isolated sites in the Sahel.

1.3.6. AUSTRALASIA. The Bureau of Meteorology (BoM) operates about 60 radars across Australia, primarily along the coast and near populated areas. The data provided to the public is more progressive than that of most countries and includes a variety of ranges and also base velocity, but the data itself is not available. New Zealand operates a modern radar network which is being expanded to 9 sites. However beam blockage by the country's mountainous terrain is a significant problem, and only the more densely populated regions are adequately covered.

REVIEW QUESTIONS

1. What might have happened if Richard Nixon had not resigned after the 1973 Watergate scandal?

2. Actually there are no review questions this in chapter. The history chapter is just meant to provide valuable context for understanding how today's weather radars were developed.

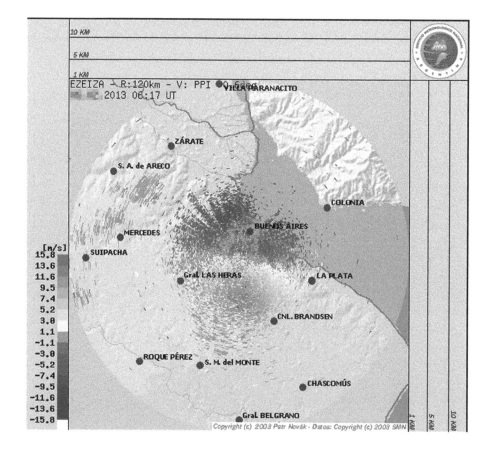

Figure 1-9. Base velocity for Buenos Aires, Argentina. This shows winds out of the north at all levels, with no directional change, giving a sort of poor man's VAD/VWP. Network images available for the public beyond coarse reflectivity are quite rare and depend on the agency's policies, funding, and public interest. This is one of the more progressive offerings, which is made all the more surprising by the fact that Latin America's weather services have traditionally been run by military agencies and to an extent are still geared primarily to meet military needs. *(Servicio Meteorologico Nacional Argentina)*

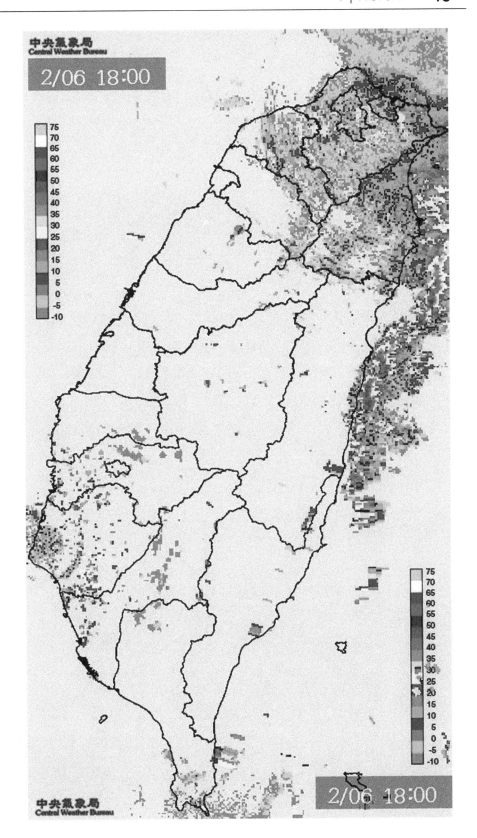

Figure 1-10. **Composite reflectivity mosaic for Taiwan**. Finding and bookmarking radar sites across the world is an excellent way to explore the workings of many interesting and unfamiliar weather systems. *(Central Weather Bureau Taiwan)*

2 Fundamentals

2 | FUNDAMENTALS

Electromagnetism is arguably one of the most ubiquitous forces of nature, not only structuring and binding the molecules that make up our bodies, our chairs, and the coffee cup on our desk, but making luxuries like electrical power, magnetism, computers, radio, and television possible. Electromagnetism, by its definition, shows is made up of electric and magnetic fields. Though the magnetic field is important to the engineering aspects of radar, it is the electric field that is important to meteorologists.

The purpose of this chapter is to lay out the underlying elements of electromagnetism and build up a foundation for the understanding of radar meteorology. This chapter is somewhat dry since it focuses on the physics of radar sets, but it should not be skipped. Without it, forecasters will not be able to understand things like the limitations of the different operating modes of radars and what causes certain precipitation particles to have unusually strong or weak intensity.

2.1. Properties of electromagnetic waves

When electromagnetic energy is radiated, the nearby electric and magnetic field are oscillated. These oscillations propagate outward, forming a series of electromagnetic waves moving through space and time. They have unique properties that distinguish them from other electromagnetic waves, and in terms of radar meteorology, affect the behavior of the radar equipment and its ability to detect different types of weather.

2.1.1. VELOCITY. As complicated as electromagnetism may be, our understanding is greatly simplified by the fact that it always moves at a constant speed: the speed of light, or c. Even though c is given as a constant, the velocity in fact depends on the medium through which the energy moves. In a vacuum, this is 300 million m s^{-1} (meters per second). The velocity can be drastically smaller in transparent materials such as glass and water, but within the Earth's atmosphere, the difference between velocity in a vacuum and velocity in the air is almost negligible. That's not to say that the differences aren't important; unusual differences in density with height, such as a temperature inversion, can indeed be sufficient to affect the behavior of the beam. For the operational meteorologist, however, the speed of light through the atmosphere can be assumed to be a constant 300 million m s^{-1} or 300,000 km s^{-1}, and this is a constant that should be memorized since it appears in many electromagnetic equations.

2.1.2. WAVELENGTH. Perhaps the most important property of electromagnetic radiation is its wavelength, often given in technical papers and journals using the Greek letter lambda (λ). Wavelength is simply the distance spanned by one full wave. This is important because transmitters, receivers, and antennas are constructed differently depending on which wavelengths they work with, and

The use of negative exponents
You may ask why an expression of meters per second is written in this book as m s^{-1} instead of m/s. You may also wonder why density is written as kg m^{-3} instead of kg/m^3. Derived quantities are written with a negative exponent form to avoid the use of inline division symbols, making the equation easier to read. It also meets the style preferred by meteorological journals and technical publications and familiarizes you with it. By flipping the sign of the exponent on the second term, we can change multiplication to division or vice versa. Take for example kg/m^3 (kilograms per cubic meter). This can be easily rewritten as kg$\div$m^3. We invert the sign of the second term to get kg$\div$m^{-3}, and invert the operator to kg$\times$m^{-3}. Since we can omit the multiplication sign in algebraic equations, this is the same thing as kg m^{-3}.

Title image
The WSR-88D radome at Oklahoma City, Oklahoma (KTLX) as seen on a cloudy day in September 2009. *(Tim Vasquez)*

Figure 2-1. The basic characteristics of a wave. Here we have highlighted one complete sine wave in white, which is embedded within a wave train of infinite length. This wave is made up of one trough, where the amplitude is negative, and one ridge, where the amplitude is positive. All of this makes up one complete cycle. Phase can be thought of as how far the wave has progressed through one complete cycle. Here, units of phase are given in units of 0 to 360 degrees, similar to one complete turn of a bicycle wheel. This concept of the wave is the basis for understanding the electromagnetic spectrum.

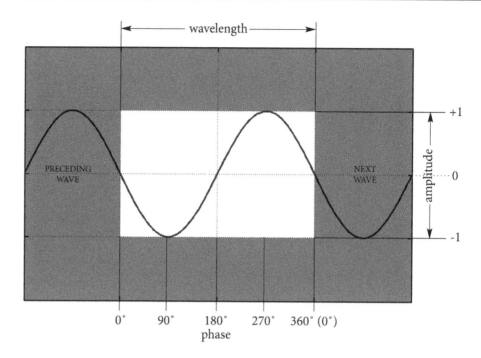

Some simple sizes
This useful table lists some of the equivalencies most frequently encountered when using WSR-88D products:

km	nm
0.25	0.13
0.5	0.27
1	0.54
2	1.1
4	2.2

the atmosphere and precipitation particles all react differently to electromagnetic energy depending on its wavelength. A good example is visible light, which has an extremely short wavelength and can be blocked by a cloud, while radio energy with relatively long wavelength can pass through all kinds of clouds and weather.

Standard weather radar equipment works specifically in the microwave band (see Table 2-1) and has a wavelength on the order of centimeters. The most common weather radar wavelength used worldwide is 5 cm, and such radar units are often referred to as "C-band" radars. Some networks, such as those in the United States and in southern China work at the 10 cm "S-band" wavelength. Such radars have a greater ability to penetrate clouds and severe weather, but the equipment and antenna are larger and more expensive.

2.1.3. FREQUENCY. Because the velocity of light in the atmosphere is about 300 million m s^{-1}, we can describe electromagnetic waves not in terms of wavelength but in the number of waves which pass a given point each second. Consider if we have a radio transmitter and we wish to produce an electromagnetic wave with a wavelength of exactly 1 meter. It follows that if we build such a device, and given that the speed of light is 300 million m s^{-1}, then this device will cause 300 million waves to pass a given point each second. So we can characterize our device as transmitting at a certain *frequency* rather than describing the wavelengths it radiates.

The unit of measure for frequency is Hz (hertz), which is simply a fancy way of saying cps (cycles per second), a unit that was actually used until the 1960s. If our hypothetical device has a frequency of 300 million waves per second, we

Table 2-1. Microwave bands. Radar systems are usually classed into bands set forth by the IEEE (Institute of Electrical and Electronics Engineers). There are other systems widely used, such as the ITU radio bands which classify radio into MF, VHF, SHF, and so forth, but this is not used in radar applications. The single-letter IEEE bands shown here are classified as microwave radio bands.

Band	Wavelength	Frequency	Use
	< 1 mm	> 300 GHz	Infrared, visible, ultraviolet, X-ray, and gamma bands
G	2 mm	150 GHz	Millimeter-wave radar
W	3 mm	100 GHz	Millimeter-wave radar, military targeting, cloud radars
V	5 mm	60 GHz	Millimeter-wave radar, cloud radars
K	1 cm	30 GHz	Cloud radars, satellite television (especially the K_u band)
X	3 cm	10 GHz	Airborne radar
C	5 cm	6 GHz	Local warning radar
S	10 cm	3 GHz	Network radar (WSR-88D, WSR-57)
L	20 cm	1.5 GHz	Air traffic control
	> 1 m	< 300 MHz	Radio bands (UHF, VHF, etc), wind profilers

More about frequency

Many amateur radio operators transmit on bands defined in terms of wavelength. One common example is the 40 meter band. How can we convert this to a radio frequency? Knowing that 1 meter equals 300 MHz, we know we need to either multiply or divide 300 MHz by 40. To resolve the confusion, it helps to picture the wavetrain coming from a transmitter. If the waves are 40 meters in length, it follows that their great length will cause a reduction in the number that pass a given point each second. So the proper solution is to divide 300 MHz by 40. This yields 7.5 MHz!

can write this as 300 million Hz, or use the kilo/mega/giga shorthand system to rewrite it as 300 MHz (megahertz). So as a rule of thumb, a *1-meter radio wavelength equals 300 MHz.*

In summary, wavelength and frequency are different expressions of the same thing. Frequency is much more useful from an engineering standpoint, and some of us encounter it every day when we tune in 97.9 MHz on our car stereo instead of 3.06 meters. Wavelength, not frequency, is by far a more important quantity in the operational aspects of radar meteorology and no further reference will be made to it in this book. Still, though, it is important to recognize the relation of frequency to wavelength, as equipment design and radio spectrum topics normally make references to frequencies.

Band names

The origin of S-band for the 10 cm band that the WSR-88D operates on originates from the term "short band". The 5 cm C-band used for warning radars gets its C letter from "compromise" between the 10 cm short band and the 3 cm X-band. The X-band, in turn, is an abbreviation for "secret band"; it was widely used by Allied radar systems during WWII.

2.1.4. AMPLITUDE. All waves have a certain amplitude, which in radar meteorology is measured as power, typically in watts. To give some idea of the problems involved in merely receiving a meaningful signal, consider that the power transmitted by a radar like the WSR-88D is just under 10^6 W (watts), while the power received by the antenna might be on the order of 10^{-12} W. Because this range of received power can be quite wide and is a miniscule fraction of the transmitted power, it is more meaningful to express it in terms of a logarithmic scale like the decibel scale, and reference it to a specific power, such as 10^{-3} W. Power in dBm equals $10 \log_{10} (P_1/P_2)$.

One factor that influences the power returned to the radar is how close the target size matches the wavelength. For example, a typical raindrop has a size of 2 mm, so a "millimeter wavelength" radar at first glance would be the ideal equipment to use. However millimeter-wavelengths suffer from strong attenuation by the atmosphere and from other droplets and cloud material, much in the same way human eyes are unable to see into a cumulus cloud.

Radar wavelengths do not necessarily need to be matched up. It is possible to sample at a longer wavelength (lower frequency), which defeats many of the problems with attenuation but requires better antennas and more sensitive receivers. The tradeoff is that short-wavelength weather radars are compact and cheap but suffer from attenuation, while long-wavelength radars are bulky, expensive, and less sensitive but have better weather penetration. In general, most conventional ground-based units operate at a wavelength of 10 cm (3 GHz), while airborne radars operate at 5 cm (6 GHz). There are also weather research radars for specialty applications that operate at much shorter wavelengths.

2.1.5. PHASE. A specific position within a wave is referred to as its phase. This can be compared to the moon's orbit around earth; if one complete rotation around the earth makes up a wave, the moon's phase is its position within that circular orbit. If two sets of waves are "in phase", this means that the ridges and troughs are perfectly synchronized with one another. The concept of phase becomes important when understanding Doppler pulse-pair processing because it influences how velocity data is extracted and explains problems like aliasing.

2.1.6. POLARIZATION. The electric field radiating from an antenna oscillates in a certain plane, much in the same way that the ocean waves which ripple onto a beach rise up and down in the vertical plane. This property is known as polarity. Conventional weather radars transmit and receive using a single, fixed polarity, typically in the horizontal plane. However by sampling a pulse volume in both the vertical and horizontal plane, it is possible to make inferences about particle shape and orientations. For example, large, falling raindrops will have a flattened shape as they fall and return much more power in the horizontal plane compared to the vertical. Measuring and comparing the electric field from two or more planes within a pulse volume is the basis for *polarimetric radar*, and will be described in more detail later in this book.

2.2. Radar energy

Now we discuss electromagnetic energy in terms of how it is harnessed in radar applications and how it interacts with meteorological targets.

2.2.1. ABSORPTION. As electromagnetic energy propagates outward from a radar site, it interacts with the molecules of the air, water vapor, water, and other matter. The electrons within the molecules are excited, which causes *absorption* of some of the energy, converting it to heat. A microwave oven is a good example of this process and furthermore illustrates how water molecules readily are excited by microwave energy, while a cardboard container remains cool. This

---

Antennas

Wavelength is important for constructing antennas. The closer an antenna length is to one-half or one-quarter of the wavelength, the more likely the antenna will "resonate" properly to where the signal can be easily detected. If the wavelength is longer, the frequency will be lower, and the antenna will be bigger!

An extreme example is very low frequency (VLF) radio, a radio band used by missile-carrying submarines due to the ability of these waves to penetrate underwater. This wavelength is 10 km. To communicate with them, very long antennas are used. The U.S. Advanced Airborne Command Post, a 747 flown by the Air Force during emergencies, contains a 5-mile long antenna reeled out at cruise altitude to facilitate communication on the VLF band!

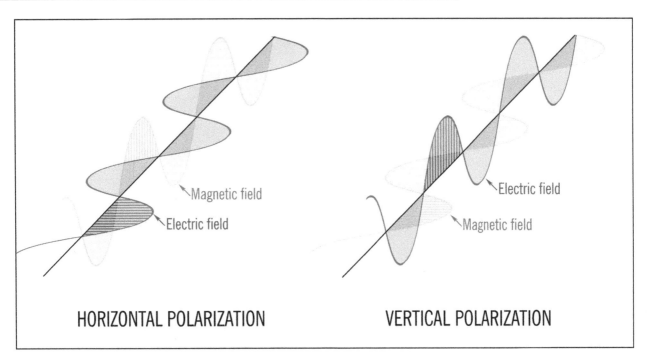

absorption is a direct conversion of electromagnetic energy to heat. The heat may be re-radiated again as electromagnetic energy, but this tends to be radiated at infrared wavelengths, not as microwave energy.

As a result, absorption by clear air and other gases, including water vapor, can actually attenuate the radar beam and reduce its ability to detect distant targets. The wavelength of the radar is critical for minimizing attenuation. Long-wavelength radars, such as the S-band systems, are less prone to this effect.

2.2.2. SCATTERING. The same process that causes electromagnetic energy to excite the electrons and produce heat also causes the molecules to behave as oscillating dipoles. They will *scatter* or "reflect" the electromagnetic energy at the same wavelength in all directions. That part of the scattered energy that returns back to the antenna is called *backscatter*, and provides the radar with meaningful information.

There are different types of scattering processes that redirect electromagnetic energy in different ways, governed largely by the size of the scatterer and the wavelength of the electromagnetic energy. For example if a particle has a diameter of approximately a fifth of the radar wavelength or less, the particle is considered to be isotropic and *Rayleigh scattering* will occur in all directions, including toward the antenna. If the particle size is greater, *Mie scattering* is the dominant mode, which *forward-scatters* most of the energy with relatively little backscatter. Even larger particles involve geometric (optical) scattering, which outside the topic of this book.

Figure 2-2. Electromagnetic polarization. An electromagnetic wave is actually made up of two fields oriented perpendicular to one another. The electric field is the component which is sensed by all operational radars, and when it oscillates horizontally with respect to the ground, it is said that the radar beam is horizontally polarized. Nearly all conventional weather radars, such as the WSR-57 and WSR-74, and pre-2011 WSR-88D use horizontal polarization exclusively. The dual-pol WSR-88D uses a clever scheme that transmits the radio energy with 45-degree polarization; i.e. imagine either of the beams pictured above rotated 45 degrees so that both are diagonal with respect to the ground. This provides a pulse with both polarization types. The receiver detects the horizontal and vertical polarization, as pictured above, of the incoming energy.

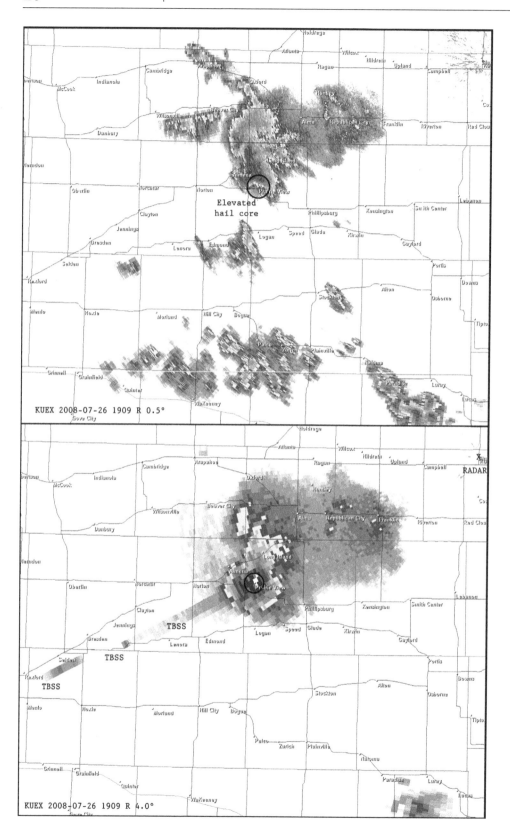

Figure 2-3. Three-body scatter spike associated with a hailstorm along the Kansas-Nebraska border on 26 July 2008. A three-body scatter spike (TBSS) is not a real meteorological feature but is an artifact caused by a bucket brigade style scattering of radar energy by highly reflective wet hail. Energy goes to the hailstone, down to the ground, back up to the hail, and back to the radar. This storm produced baseball-sized hail.

For example, the WSR-88D has a constant wavelength of 10 cm. This means that 2 cm is the threshold diameter, where smaller particles involve Rayleigh scattering and larger particles involve Mie scattering. With 3 cm airborne radars, this threshold value drops to 0.6 cm. The reason why radar meteorologists are concerned with whether Rayleigh scattering or Mie scattering is taking place is because the simple reflectivity-rainfall relationship is based primarily upon Rayleigh scattering. Once Mie scattering is occurring, this relationship breaks down and a much more complicated set of equations must be used. There are more assumptions involved and it becomes very difficult to make accurate precipitation estimates.

Scattering behaviors also change when different types of radar equipment are used. For example, the WSR-88D operates with a wavelength of 10 cm while a cheaper C-band radar has a wavelength of 5 cm. This means that for any given pulse volume there is a greater tendency for Mie scattering to become involved when the C-band radar is used. Therefore, shorter wavelength radars have more difficulty estimating precipitation totals since the reflectivity-rainfall relationship breaks down with Mie scattering.

When an extremely efficient scatterer is present, particularly in the case of large wet hail, energy may reflect out of the pulse volume to other nearby scatterers or to the ground, reflect back into the pulse volume, then reflect again back to the antenna. This energy arrives at a slightly later time than the original backscatter, and since the radar assumes this backscatter is occurring further along the radial, it will appear as a spike extending outward radially from the intense core. It is known as a *three-body scatter spike* (TBSS). In hailstorm events it can cause significant artifacts on all base products.

2.2.3. REFRACTION. Due to the decrease in density with height caused by the transition from warm, humid air at the ground to cold, dry air aloft, the speed of light increases slightly with height, causing an effect known as refraction. As a result, the radar beam is never a straight line but bends downward, having a curvature of about one-third that of the Earth. Even if the antenna was perfectly horizontal and beam blockage was not a problem, the radar does have some ability to see over the horizon. In some instances, unusual atmospheric conditions may cause the radar tilt to be higher or lower than expected, causing anomalous refraction. These artifacts are described at length in the next chapter.

2.2.4. SIDELOBES. Not all of the electromagnetic radar energy can be directed radially. Some of the energy is diffracted off the edge of the anntena, resulting in many spikes extending from the radar dish at angles to the radar beam. These are called *sidelobes*, and there is no type of antenna design that can completely get rid of them. Sidelobes can travel several miles to various objects and return back to the radar, producing false echoes that are mapped as being along the radar beam but which do not actually exist at that location. This is one of the sources

Sky scattering
Mie scattering and Rayleigh scattering can actually be understood quite intuitively using the optic spectrum. Assuming a constant composition of the atmosphere (D), small wavelengths (λ) toward the blue end of the spectrum are more likely to produce Rayleigh scattering since this lowers the result of D/λ, so blue colors have a greater tendency to disperse equally in a unidirectional fashion. Red colors involve long wavelengths, producing lesser amounts of Rayleigh scattering, so red colors tend to forward-scatter in a directional rather than a random fashion. As a result, blue is the predominant color!

Particle sizes

Type	Diameter (cm)
Cloud droplet	0.001
Drizzle	0.01 - 0.15
Ice crystals	0.01 - 0.15
Raindrops	0.01 - 0.5
Large crystals	0.15 - 0.5
Dry snow	0.01 - 0.5
Wet snow	0.5 - 3
Graupel	0.1 - 0.5
Small hail	0.5 - 0.75
Moderate hail	0.75 - 2
Large hail	2 - 5

of "ground clutter" and explains why it may appear on radar imagery even if the beam is definitely known to be pointed well above the ground.

With the WSR-88D, sidelobes are most likely to occur at a 5 to 10 degree angle from the beam. They may graze the edges of storms when the antenna is pointing into clear air, creating an apparent tangential widening of the radar echo.

2.3. Radar energy

Those who are well versed in science understand that radar is essentially an echo location process, in the same way that bats will make ultrasonic chirping noises and listen to the echo's direction and timing to form a picture of its environment and where there is prey. In the same way, radars transmit a very short electromagnetic *pulse*, a short burst of energy on the order of microseconds. This is focused into a very narrow beam so that the direction and distance from any echo can be accurately determined. There is a *listening period* in which the radar receiver detects the echoes of this pulse. The listening period is very long and comprises close to 99.8% of the radar's duty cycle. Since it takes hundreds or thousands of pulses to build up a useful radar picture, there is a continuous cycle of brief pulses and long silent periods. The characteristics of this cycle and the properties of the beam itself have significant effects on the performance of the radar, and are explained below.

2.3.1. EQUIVALENT REFLECTIVITY. It is important to remember that the radar receiver expresses the amplitude of backscattered energy in terms of power received (P_r). As described earlier, the energy WSR-88D might emit 10^6 watts of power but only receive 10^{-14} watts of backscattered power from a rain shaft, a difference of many, many magnitudes. The ability for weather phenomena to backscatter radiation depends not only on the particle size and composition, but also range to the target. Just as a candle moving away in the dark rapidly gets dimmer with increasing distance, the power from electromagnetic energy diminishes with distance due to the inverse square law, written as $1/r^2$, or simply as r^{-2}.

Since the reflectivity will be much stronger from close targets and very weak from distant ones, radar processors use the equivalent reflectivity (Z_e) equation to correct the backscattered power in a given volume. This equation uses P_r for returned power, R for range, C_r for radar constant which is specific for that radar hardware, and L_a for the attenuation factor.

Equivalent reflectivity: $Z_e = (P_r r^2) / (C_r L_a)$ (2.1)

Since this equation is a simple ratio, enlarging any of the values on the left side of the divisor will yield a higher Z_e value. The values on the right can be considered constants. Perhaps the most important term is r^2, which corrects for

Figure 2-4. Summer thunderstorm near Red Hook, New York on 17 June 2011, showing the 0.5° reflectivity (right) and visual appearance (below), looking southwest, at the exact same instant. The storm produced a weak microburst with wind gusts of 40 mph, as evidenced by the "rain foot" on the right side of the photo and a brief gustnado. Velocity products showed no velocities of more than 5 kt in the storm, but this is a range-elevation problem: the 0.5° tilt intersects the storm 37 nm away and at a height of 2700 ft AGL, so it is largely overshooting volumes below the cloud. *(Tim Vasquez)*

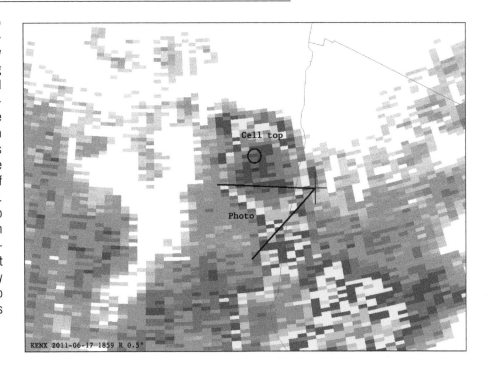

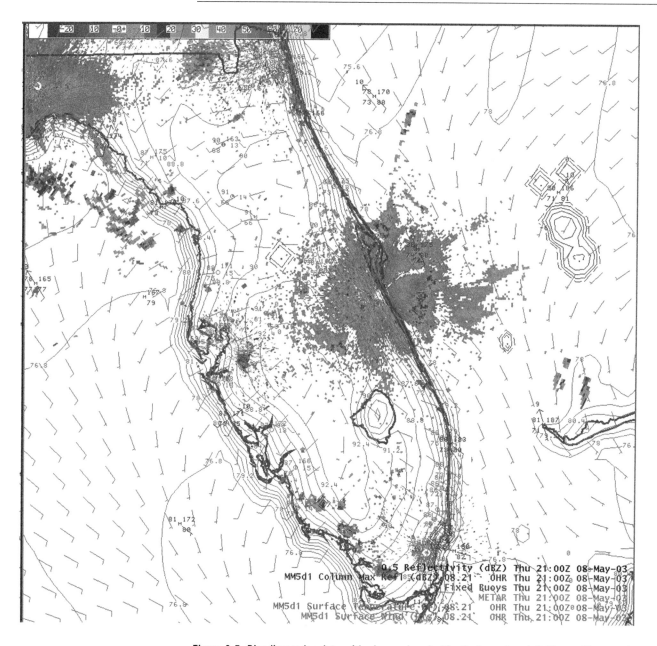

Figure 2-5. **Blending radar data with observed and objectively analyzed fields** provides an additional level of insight on the weather situation, because radar alone does not tell the whole forecast picture. Blending is not always available, so the forecaster must diagnose different products individually and assemble them mentally. The stack of product legends at the bottom right shows the distinctive appearance of an AWIPS plot, AWIPS being the primary forecasting computer system used at National Weather Service offices.

energy loss due to the inverse square law. So the equivalent reflectivity equation mainly serves to remove the influence of distance on reflectivity. However as a special note it does assume this radiation comes from Rayleigh scattering, the dominant scattering process in rain drops, not Mie scattering, which is directional and occurs with ice.

This result, Z_e, or *radar reflectivity factor*, is expressed as $mm^6\ m^{-3}$, which relates the number of drops and their size per cubic meter. The mm^6 portion of this factor uses a sixth power because a doubling of drop diameter increases reflectivity 2^6, i.e. 64 times, while a tenfold diameter increase (10^6) increase results in a millionfold reflectivity increase.

We almost have our familiar expression of "intensity" except for one problem. The values of Z_e may range anywhere from $10\ mm^6\ m^{-3}$ in light snow situations to $10^6\ mm^6\ m^{-3}$ in hail. This is a range from ten to a million. Is more useful to refine equivalent reflectivity values by applying a logarithmic decibel scale. This expresses Z_e in terms of decibels of reflectivity instead of $mm^6\ m^{-3}$:

Decibels of equivalent reflectivity: $dBZ = 10\ \log_{10} Z_e$ (2.2)

This number is widely used in radar meteorology and may be described simply as *intensity*. An intensity of of 30 dBZ normally corresponds with rain, while 60 dBZ is associated with hail or very heavy rain.

2.3.2. Pulse length. The simplest books on radar teaches that radar signals are made up of pulses, much like the ultrasonic pulses used by bats for locating prey. The pulse length is often described in terms of time, normally in microseconds (μs). Using the speed of light, we can also express the pulse length in terms of distance from the start of the pulse to its end, remembering that it travels at the speed of light. Radar engineers are normally interested in the time expression, while forecasters are more interested in its length. A typical WSR-88D pulse length is 1.57 μs, which equals 470 meters.

2.3.3. Pulse Repetition Frequency. Radar energy is transmitted with not just one pulse but a series of them. The interval between the start of one pulse and the start of the next pulse is referred to as the *pulse repetition time* (PRT) or *pulse interval*. This time expression is of interest to radar engineers, but for operational meteorologists it is actually more meaningful to express it as "pulses per second", in other words, as a frequency. This is known as *pulse repetition frequency* (PRF). This has nothing to do with radio frequency but is simply how many pulses occur in one second. We can go one step further and use the speed of light to express PRF or PRT as a distance. A radar, such as the WSR-88D, operating at a PRF of 1304 will show 0.8 ms for PRT, 1304 Hz for PRF, and 230 km for PRF distance.

Decibels of reflectivity
For those unfamiliar with what the term $10\ \log_{10} Z_e$ means, here are some examples of radar reflectivity factor converted into a decibel of intensity:.

Z_e ($mm^6\ m^{-3}$)	Z_e (dBZ)
$10\ \log_{10}$ (1)	0 dBZ
$10\ \log_{10}$ (10)	10 dBZ
$10\ \log_{10}$ (100)	20 dBZ
$10\ \log_{10}$ (1000)	30 dBZ
$10\ \log_{10}$ (10,000)	40 dBZ
$10\ \log_{10}$ (100,000)	50 dBZ
$10\ \log_{10}$ (1 million)	60 dBZ
$10\ \log_{10}$ (10 million)	70 dBZ

A brief VCP overview
This is a quick reference summary of WSR-88D VCPs. Terms here include MPDA (multi-PRF detection algorithm).

VCP 11. The basic precipitation mode algorithm.

VCP 21. Similar to VCP 11 but uses 9 elevation angles instead of 14.

VCP 121. Same as VCP 21, but uses MPDA to reduce range/velocity aliasing.

VCP 211/212/221. Same as VCP 11, 12, and 21, respectivity, but these use the Sachidananda-Zrnic split-cut algorithm to improve unambiguous range.

VCP 31. The basic clear air mode algorithm.

VCP 32. Same as VCP 31 but offers a short pulse to provide higher unambiguous velocity.

F

2.3.4. MAXIMUM UNAMBIGUOUS RANGE. While the physical range of the radar is largely dictated by how powerful the transmitter and how sensitive the receiver are, the *maximum unambiguous range*, of a radar is a function of its pulse repetition frequency. This is rather intuitive, because if the radar emits more pulses while backscattered energy is still on its way back to the antenna, this will corrupt the resulting radar imagery and result in an artifact known as a *second-trip echo*. This problem is known as range folding.

To describe this in more detail, the maximum practical range of a radar can be simply written as:

$$R_{MAX} = c / (2 \times \text{PRF}) \tag{2.3}$$

This can be understood rather intuitively. First, let's go ahead and solve it using a typical radar PRF of 1000 Hz. Since c is 300,000 km, we get an R_{MAX} of 150 km. Now, let's use mental imagery to describe the radar pulse's journey. If a PRF is 1000 Hz, this means that the PRT is 0.001 second. Multiplying this PRT by 300,000 km s^{-1} (the speed of light) this means a pulse will only travel 300 km before the next pulse is emitted.

But this is just a trip in one direction! Radar energy, in order to be meaningful, must go from the transmitter to the target and then bounce back to the radar. A radar signal that strikes a target and returns to the radar, having traveled a total of 300 km, will only have traveled 150 km and then another 150 km back to the radar again, and that 150 km distance matches the limit we calculated above.

Obviously it would be a simple matter to just increase the PRF to improve the maximum resolution of the radar, but doing so means that the radar antenna must move slower due to the long listening times, and in order to build a picture. The choice is either to accept a lower PRF and reduce the radar range or accept a higher PRF at the cost of slower scans. Performing sector scans can be one way of overcoming the slowness of product output cycles by limiting the area of detection to a certain range of azimuths, but in practice this is done only with research radars.

2.4. The radar beam

The radar beam is often thought of a straight, laser-like line, but it is actually conical and has appreciable width. Some of the early WWII radars had a beam width of over 20 degrees! The WSR-88D, by contrast, has a beam width of just under 1 degree. This has important consequences, because even at 1 degree a pulse 100 miles from the radar will be two miles in height and width. As a result, features in a large volume like this that are much smaller than the beam width tend to show a weaker power return and might give false indications of intensity and velocity.

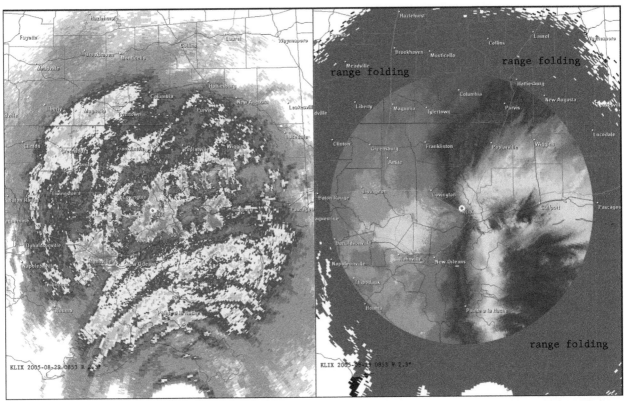

Figure 2-6. Range folding of radar data, as seen in an example of Hurricane Katrina. The reflectivity data (left) has greater immunity to range folding because the signal processor makes a pass for lower tilt reflectivity at low PRFs. However velocity data (right),, which requires higher PRF to minimize aliasing, is more vulnerable. Here it clearly shows range folding beyond 63 nm, which at this range, wavelength, and PRF consitutes RMAX during the velocity sampling.

Another misconception is that the radar energy has definite angular width. In other words, if we use a theater stage spotlight to represent the radar beam, we would not see a circle of light on the stage with hard edges marking the boundary between light and dark, but more of a soft, fuzzy transition. Since this means there are not definite edges with which we can measure the beamwidth, this figure must be expressed in terms of half-power points, the points on either side of the beam where the power is 50% of that of the maximum power in the center. This angle between the two half-power points constitutes the radar's *physical beamwidth*.

The volume that falls between the half-power points at any range is referred to as the *illuminated volume* and the volume that contains the radar pulse at any given instant is called the *pulse volume*. This pulse volume is sometimes referred to as a "gate" or more properly as a "bin" and is the most granular level of digital radar output.

2.4.1. Coordinate Systems. Any given point on earth is usually described by its latitude, longitude, and elevation. This set of numbers is called a coordinate system. The coordinate system for an individual radar is azimuth, elevation, and range. Also an axis aligned parallel to the radar beam is referred to as *radial*, so if we are measuring the component of velocity away from the radar dish, this is

Rule of thumb
For a one-degree beamwidth radar like the WSR-88D, the beam has a diameter of 1 mile for every 60 nm (100 km) of distance.

A-scope
The A-scope was used in the first- and second-generation radars and is almost never encountered anymore. It was a long-phosphor CRT display that showed increasing range along a radial from left to right and intensity from bottom to top. This was used when the radar dish was under manual control. Using a dial, the radar operator moved an electronic notch to targets seen on the A-scope and read the distance from a mechanical display to obtain the exact range to intense storm cores and other features.

referred to as radial velocity. Axes that are perpendicular to the radial are called *tangential*.

The term *azimuth* simply refers to which direction the beam is pointing relative to true north. Normally the radar makes an entire sweep of all azimuths in about one minute. This type of sweep is referred to as a surveillance scan. In research meteorology, a radar might only sweep certain directions for physical or technical reasons, such as to get better temporal resolution. This type of scan is called a sector scan.

In the context of radar meteorology, *elevation* refers to the upward tilt of the beam from the horizontal plane. Sometimes *tilt* is preferred to avoid confusion with the idea of altitudes, but both elevation and tilt mean the same thing in radar meteorology. Normally the radar beam is tilted slightly (about 0.5° upward) to maximize the range without introducing beam blockage from objects on the ground. Before the advent of modern Doppler radars, network radars were operated continuously at an elevation of about 0.5° except at times when storms were being manually analyzed by the radar technician. Today's automated radars are programmed to continuously scan many different elevations. What we see on basic radar products such as those on television and public weather websites consists of either the lowest elevation or a composite image of all elevations.

Finally, there is the term *range*. While in aviation range might refer to the maximum flying distance of a plane on a single load of fuel, the maximum usable distance of a radar is specifically referred to as maximum range. The term range, by itself, is any expression of radial distance, that is, along the beam between the radar antenna and a given point in the atmosphere. So if a storm is 60 miles from the radar site, it is described as being located at a range of 60 miles.

2.4.2. RESOLUTION. The term resolution refers to the ability of the radar to detect and distinguish small-scale features. Naturally the goal is to minimize the size of the pulse volume, but the ability to do this may be affected by insufficient resolution.

Angular resolution refers to the ability of the radar to distinguish small-scale features laterally, or *perpendicular* to the radar beam. One of the factors that dictates angular resolution is the antenna design. Antennas that provide a narrower beamwidth are more expensive to construct. The available processing algorithms also affect angular resolution. For example, though the WSR-88D beamwidth is about 1 degree, it is possible to use oversampling techniques to achieve an effective angular resolution or "virtual beamwidth" of 0.5°. This is described in more detail in the next section, Volumetric Scanning.

Radial resolution refers to the ability of the radar to distinguish small-scale features longitudinally, or *along* the radar beam. This is mostly dictated by the pulse length. If a radar uses a long pulse length, more power is fed into the pulse and the radar needs less sensitivity to detect the echoes, but at the same time

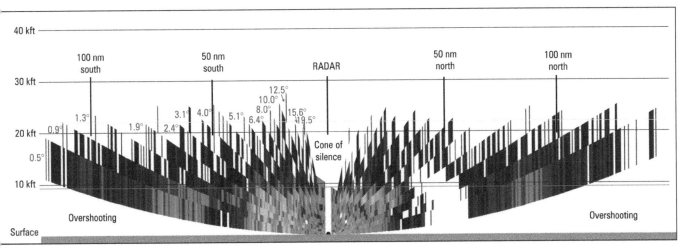

Figure 2-7. Vertical cross section across a radar volume-during a stratiform rain event in Pennsylvania. This clearly shows the conical property of the radar beam, made up of a stack of different tilts (labeled at left in degrees). Also seen here is the dramatic beam widening with distance, the outermost pixels ballooning to 2 miles in diameter. The cone of silence is the region above the radar not sampled by the radar beam. In the case of the WSR-88D this is all elevations past 19.5°.

the large radial length of the pulse means that power is averaged across a wider distance, and as a result the radial resolution of the radar is degraded. Pulse length can be decreased to improve the radial resolution of the radar, but a more powerful transmitter or a more sensitive antenna and receiver must be used. Oversampling algorithms can also compensate and improve the effective radial resolution of the radar.

2.4.3. BEAM SPREADING. A common misperception is that any given "gate" of radar data represents the highest reflectivity or velocity at that point. What is displayed is only an average of that property throughout the pulse volume. So if a hail core was able to show 80 dBZ if it could fill the entire pulse volume, it might only show 40 or 50 dBZ if it occupies only a small part of that volume. Beam spreading at significant ranges and the existance of small features and circulations at these ranges intensify the effect.

2.4.4. THE CONICAL NATURE OF A TILT. Because of the fact that the radar beam does not curve as strongly as the earth and the fact that the earth falls away with increasing distance due to its own curvature, all radar beams increase in height with distance. What this means is that the illuminated part of any 360-degree azimuth sweep, regardless of tilt, will always paint a *conical* shape. Perhaps the most important skill in radar analysis is being able to properly visualize horizontal radar graphics as conical, not flat, with targets near the radar being located at low altitudes and those around the edges at high altitudes. As the radar tilts upward from 0°, this effect becomes substantially greater.

2.4.5. BEAM BLOCKAGE. Although weather agencies try their best to site the radars where they have a clear view of the sky, there are usually a variety of

> The WSR-88D transmitter is considered to be a coherent radar, which means that the phase (timing) of the waves produced by the RDA is maintained with great precision in order to properly extract velocity information.

obstructions that block part of the beam near the ground or obscure the entire beam. Mountains, hills, buildings, trees, and antenna farms are the normal beam blockage culprits. This reduces the energy returned to the radar from targets beyond, or may even eliminate the energy altogether. For the WSR-88D, blockage maps for each radar site are published at the NCDC (National Climatic Data Center) website. It is imperative for forecasters to become familiar with the ones in their local area, particularly in mountainous parts of the country, so that when detail is lost while analyzing a storm, the possibility of beam blockage might be recognized.

2.4.6. DISTANT RANGES. A common assumption is that the radar detects all precipitation within range of the radar. This fails when the precipitation or weather feature is occurring below the lowest available tilt. At the range of 124 nm which marks the outer limit of most WSR-88D products, the 0.5° tilt is at a height of about 17,000 ft. Low stratiform precipitation, low-topped thunderstorms, mesocyclones, tornadoes, fronts, and other boundaries are easily missed, regardless of how precise or powerful the radar hardware might be. Not only will their signatures be missing from reflectivity products, but precipitation estimates cannot be accurately determined to due the missing data. This is a physical limitation and the only practical solution is fielding more radars.

Because of this and the beam spreading issues, proper analysis of distant thunderstorms becomes increasingly difficult beyond roughly 50 to 60 miles. Forecasters should always double-check to make sure there is not a closer radar within range of distant targets.

> **What to do at long distances**
> Forecasters should have a working knowledge of which areas in their region are distant from the radar network. With the existing WSR-88D network the most prominent examples are found in far southeast Montana, central South Dakota, northeast Missouri, far south central Missouri, far southeastern Arkansas, the Red River region of northeast Texas, and the area where the borders of Colorado, New Mexico, and Oklahoma meet. Assessment of storm severity is difficult in such areas and tornado detection is unreliable. Secondary sources such as spotter reports, chaser webcasts, lightning detection, surface data, and even satellite data should all be used heavily when weather events are occurring in these areas. It may also be helpful to compare distant radar data from multiple sites. A map is provided in the appendix which shows radar quality across the United States.

2.4.7. CLOSE RANGES. At distances very close to the radar, roughly within 10 to 20 miles, weather returns are degraded by "ground clutter" and sidelobes. At such distances it may be useful to use higher tilts than normal to interpret radar data. With severe storms within 30 miles of the radar, elevations of about 1° to 2° often yields more useful data than the normal 0.5° tilt, particularly with velocity data, and the beam is more likely to pass within the base of the thunderstorm. The lowest tilt should still be monitored to detect circulations, boundaries, and precipitation developing near the surface.

There are also physical limits to how high the radar antenna can tilt and the volume scan's highest design elevation. For example with the WSR-88D the maximum available tilt is 19.5°, corresponding to an elevation of only 11,000 ft just 5 miles from the radar. Anything above this maximum tilt falls within a zone called the *cone of silence*, with the cone centered on the zenith and its edges lying on the surface of the highest available tilt. No tilts are available above 19.5° because the WSR-88D has only a 1° beamwidth and it would take about 30 tilts to sample all elevations up to 90°. There is no way to do this without adding tremendous time to the volume scan or introducing large data gaps at these higher elevations. Therefore targets within the cone of silence cannot be sampled meaningfully and the region is omitted as a design consideration.

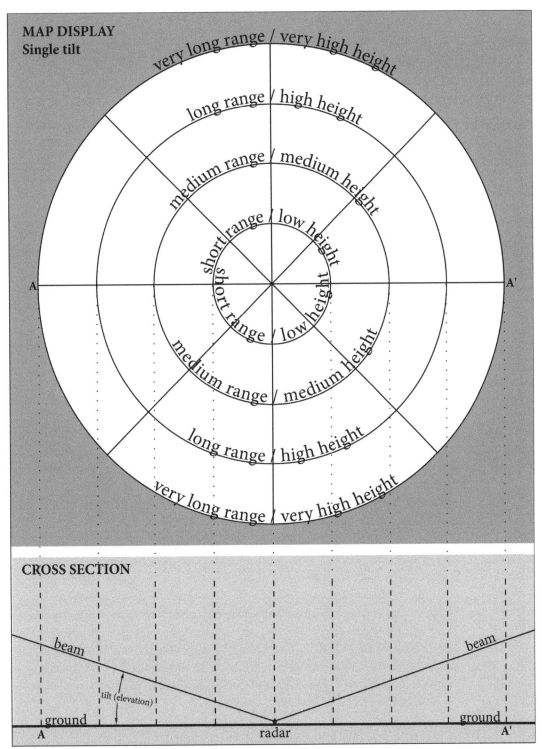

Figure 2-8 The conical nature of a radar tilt (left, facing page) is diagrammed in detail. This is one of the single most important concepts for interpreting radar patterns and understanding some of the problems associated with derived products. The great majority of the concentric or symmetric patterns centered on the radar site can be explained by this diagram. It also forms the framework for all general use of single-tilt radar products. Even when examining very localized areas, the forecaster will always be aware of what the distance of this area is to the radar site, and in turn, whether the area is close to the ground or high in the atmosphere. For purposes of simplicity, the Earth's curvature and refraction of the radar beam are neglected here.

2.5. Volumetric scanning

Modern radars treat the entire atmosphere sampled as a *volume*. It may take 5 to 10 minutes to collect all of the data to form one complete volume, which is why Internet radar products and desktop radar viewers take several minutes to refresh with new data. Each volume consists of scans at specific elevations, each one known as a *tilt*. Each of these tilts consists of data along many different azimuths, each of these referred to as a *radial*. And within each radial, we reach the limit of granularity of the radar, where the radial is broken into individual points of data. Each point is referred to as a *bin*.

2.5.1. SCANNING STRATEGIES. Volumetric scanning radars use a number of scan strategies to overcome the requirements of update frequency, problems associated with PRF selection, and what type of weather phenomenon is being analyzed. As a result, a scan strategy will use specific sets of tilts, specific azimuth scan rates, specific PRF rates, and other specific configurations. The WSR-88D has its own system of scan strategies, while volumetric radars operated by other countries will use their own systems of scanning strategies.

2.5.2. VOLUME COVERAGE PATTERN (VCP). The scanning strategy schemes used by the United States' WSR-88D network are known as volume coverage patterns (VCPs), each one identified by a specific number. When the radar was first rolled out in the early 1990s there were four of these, VCP 11 and 21 for precipitation and VCP 31 and 32 for clear air operation. During the 2000s many new VCPs were introduced which optimized the radar and reduced the effects of range and velocity aliasing. These are listed briefly in the sidebar and in more detail in the appendix.

2.5.3. BASE PRODUCTS. Base products are the core products produced by the radar's signal processor upon which *all* other products are based. The three base products of most volumetric radars, including the WSR-88D, are reflectivity, velocity, and spectrum width. The first two may be called base reflectivity and base velocity to emphasize that they are not derived products. Base reflectivity is sometimes erroneously described as the lowest tilt of the radar, when in fact this is simply a single tilt at any elevation coming straight from the radar processor without any change. Polarimetric radars have additional base products which are used to create derived polarimetric products.

2.5.4. DATA. Though many radar products are available on the Internet in the form of *images*, the power to fully manipulate and explore the radar information requires access to the *data* itself. In the United States, WSR-88D data is distributed in the form of Level II and Level III messages, which are publicly available on NOAA servers and datastreams and are the foundation for a wide variety of data viewers. In other countries, raw data from weather radars is not

---

A road to transparency

During the 1990s, the cost of transmitting data was considerably higher than it is today. Foreseeing problems with having to manage a large number of interfaces at each RPG if the general public were allowed access, the NEXRAD program instead allowed four vendors, Kavouras, WSI, Unisys, and Alden, to access the RPG at each site and take on the burden of distributing data to the public. The vendors each paid about $130,000 per year for the data. This arrangement was known as NIDS (NEXRAD Information Dissemination Service). Since the vendors recouped their expenses by selling the realtime data as a premium product, it was difficult for hobbyists and the general public to access good NEXRAD imagery since only degraded products were offered for free. Effective January 1, 2001, the NIDS agreement ended and the National Weather Service assumed responsibility for distributing the data over available networks such as NOAAPORT and Internet servers. This led to a proliferation of display software that continues to this day on mobile devices like the iPhone.

available to the general public. A wide variety of proprietary data formats are common, but many weather agencies are working toward adopting the BUFR format, prescribed by the World Meteorological Organization, as well as HDF5, a data storage format widely used by many scientific institutions. Some data is also exchanged in NetCDF format, another data storage standard.

2.5.5. BEAMWIDTH OPTIMIZATION. Weather radars have a physical beamwidth dictated largely by the geometry of the radar dish. In the case of the WSR-88D, half of the power is concentrated in a 1.0° beam, so the radar is said to have a one-degree beamwidth. Naturally we can construct a volume scan that has 360 radials measuring 1 degree in width each.

However this introduces some problems. Consider a volume scan whose radials are made up as follows: 0.5° (0.0°..1.0°), 1.5° (1.0°..2.0°), 2.5° (2.0°..3.0°), etc. If the radar simply emitted one pulse at an antenna centerline of 0.5°, the next at 0.5°, the next one at 2.5°, and so on, each power return will be assigned to the correct radial. However this wastes time and it is advantageous for the WSR-88D to transmit about 11 pulses per degree while the antenna rotates. This means that the 0.5° radial will fall within a half power point from the time the antenna centerline is at 359.5°, where a half-power point is at 0.0° and grazes the edge of the radial in question, until the antenna rotates to 1.5°, where the other half-power point is at 1.0° and grazes the other side of the radial. Since the radial is illuminated by any antenna azimuth from 359.5 to 1.5°, the radar has an *effective beamwidth* of 2.0°. In practice, however, the WSR-88D's effective beamwidth is 1.4°.

This effective beamwidth can be narrowed to fit a radial by *data windowing*. This uses a weighting function, where the pulses used to construct data for a radial are weighted heavily if they fall entirely within the radial and more weakly if they fall largely outside the radial. The data with higher weighting scores are used to construct the base quantities of reflectivity, velocity, and spectrum width for that radial.

This technique can also be used for high-resolution radials like the WSR-88D super resolution products. The problem is that increasing the resolution beyond the antenna's physical beamwidth introduces mathematical

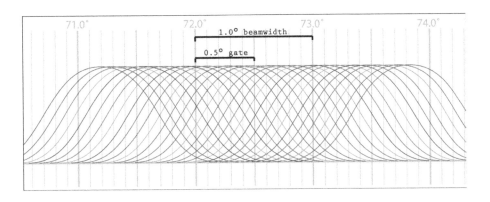

Figure 2-9. An explanation of oversampling, as shown here for radials near 72 degrees of azimuth. Each sine wave represents the potential for the WSR-88D to receive backscattered power from an imaginary target at a given azimuth, assuming the beam's centerline is at the peak of the wave; for example if the beam is centered at 71.2°, a target at 71.7° will only reflect about half the power. It can be clearly seen that a very small target might be sampled by dozens of radials. Furthermore, its exact location might be narrowed down based on which exact radials detected the energy and how the power varied from radial to radial.

errors, seem as noise and graininess. This is enough of a problem to where many of the WSR-88D algorithms actually use the standard 1.0° radials, not the super resolution radials.

2.5.6. CLUTTER REDUCTION. As mentioned earlier, ground clutter appears to be close to the radar, within 5 to 20 miles, and is produced mostly by interactions between antenna sidelobes and terrestrial objects. There are two types of ground clutter: semi-permanent clutter, such as mountains and buildings, and transient clutter, such as airplanes and vehicles. Clutter can be ignored by human forecasters, and this was in fact what was done with conventional radars through the early 1990s. However with the arrival of volumetric radars and the proliferation of algorithms, there is not only the means to reduce clutter but the need to do so to prevent corruption of algorithm output. The WSR-88D radar uses an algorithm known as the clutter mitigation decision algorithm (CMD) to reduce these effects.

2.6. Visualization

The "old fashioned" radar displays often seen in movies made in the 1960s and 1970s were marvels in themselves. Memory was too expensive to allow the display to continuously "paint" the patterns of radar echoes in areas that the radar was not sampling, so the cathode ray tube displays were coated with long-persistence phosphor, which would capture the energy from the radar sweep on the CRT and cause those painted areas to glow faintly for a short time. The glow was so faint that radar operators had to work in darkened areas, and black curtains were a staple of many Weather Bureau offices.

A revolution occurred with digital display of data on television sets and computer monitors. Though the basic PPI (horizontal) and RHI (vertical) display types no longer exist in their original form due to the advent of multipurpose digital displays, they are still regularly mentioned in journal papers and by older forecasters and remain an important part of the meteorological lexicon.

2.6.1. PLAN VIEW. The term "plan view" refers to any view looking directly downward, displaying the data as if it were plotted on a map. Normally the X-axis points eastward and the Y-axis northward. One noteworthy type of plan view is constructed by a plan position indicator, or PPI, the primary type of graphics display for first- and second-generation radars.

2.6.2. CROSS SECTION. A cross section is a view looking not downward but sideways. The X-axis points to an arbitrary azimuth (direction) and the Y-axis points upward in the atmosphere.

First and second generation radars had a separate phosphor display in which the radar antenna was locked on a certain azimuth and the antenna tilted up and

---

Universal Coordinated Time

All forecasters must be familiar with UTC (Universal Coordinated Time). This is simply the standard time in London, England, ignoring any correction for Summer Time. All meteorological data, including radar data, is expressed in UTC. When the U.S. is off daylight saving time, Pacific Time adds 8 hours, Mountain Time adds 7 hours, Central Time adds 6 hours, and Eastern Time adds 5 hours to obtain UTC. During daylight saving time in the U.S., the value of each number is decreased by one (so the corrections are 7, 6, 5, and 4). For those not used to this, don't worry; it will soon become second nature!

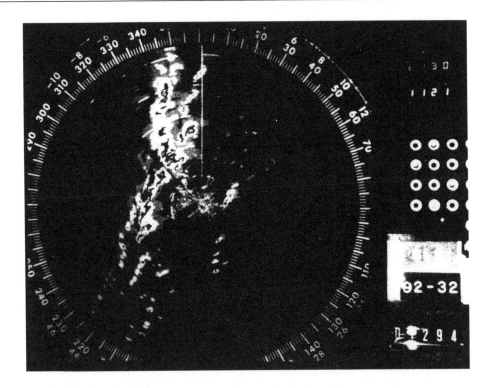

Figure 2-10a. Plan position indicator (PPI) of a conventional WSR-57 radar. This type of display forms the basis for modern radar products, nearly all of which use the plan view. The "surface" used by a radar plan view is never horizontal, i.e. the echoes shown are not for a fixed level above ground or sea level, but is conical, with distant echoes at high altitudes and close echoes at low altitudes. This is due to the conical nature of a radar beam. *(National Weather Service)*

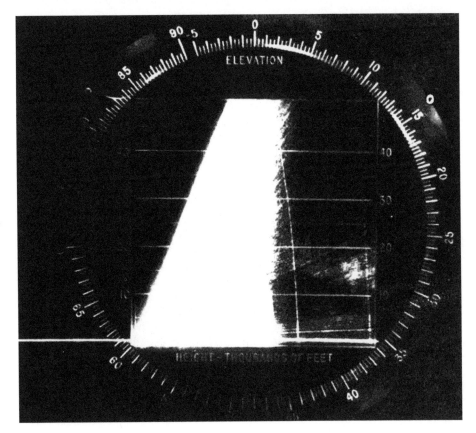

Figure 2-10b. Range height indicator (RHI) of a conventional radar. This scope was ignored until the radar operator had manual control of the antenna. The operator would then point the dish at a desired azimuth and tilt the antenna up and down, painting the images seen on this scope. This would allow storm tops to be measured and help the operator identify severe weather signatures like WERs and BW-ERs. The same RHI display can be recreated in programs like the basic version of GRLevel2, which offers a Range Height Indicator display. However this is limited to radials along the radar beam, and it is much more useful to use the regular cross section capabilities, which can be freely created along any axis.

Bins and gates
Though bins and gates are often thought to be synonymous, bins refer to the actual samples in the base data. On the other hand, gates generally refer to pixels on radar images, which may actually be displayed at a different resolution. Gates, according to some, is also the person who says Microsoft will buy you a Disneyworld vacation if you forward a certain chain e-mail to others. For those who still haven't caught up with the hoaxes of 10 years ago, the answer is no, he won't!

down. This produced a cross-section along that azimuth, with range increasing from left to right and height increasing from bottom to top. This is known as the range-height indicator (RHI). Ironically, an RHI display built from volumetric scans of third-generation and later radars does not have the quality of RHI scans from the earlier units because the volumetric scanning strategies introduce numerous elevation gaps and a prominent cone of silence.

Volumetric display software packages are not locked to the beam radial but can be constructed anywhere within the radar volume. This is usually tone by "dragging" the cursor from one point to another or clicking in two places on a plan view, then viewing the resulting cross section along that axis in a separate window. Due to the inherent problem of gaps between tilts in a volumetric scan, the vertical resolution is not as good as older radars like the WSR-57, and the software display packages may either show the gaps or will attempt to interpolate within them. The cone of silence near the radar will be highly prominent on all volumetric cross sections.

2.6.3. MULTIPLE WINDOWS. One important type of display offered with the very first WSR-88D workstations and provided nowadays in many radar display software packages is the ability to provide two or more windows, each containing different products from a volume scan, and with the cursor displayed on all of the windows in the correct 3-D space. This is an important capability that helps the forecaster evaluate continuity of a feature across different types of products or different tilts and better visualize what is happening in an area of interest.

2.6.4. THREE-DIMENSIONAL (3-D) DISPLAYS. Using volumetric scans it is possible for computers to display a three-dimensional representation of a storm, though the useful of this is currently limited due to the inherent limitations of "simulating" a three-dimensional scene on a two-dimensional display and the fact that a complex process like a thunderstorm does not have any sort of simple surface lending itself to 3-D visualization. Two techniques are typically used for 3-D representations: one in which various levels are semi-transparent (lit volume), and one in which a specific surface, such as reflectivity of a specific power, are displayed (isosurface). The isosurface method is generally most used since the favored technique for it is to probe rain-filled parts of the cloud for patterns and structures.

2.6.5. BIT LEVELS. In the early days of digital radar, it was expensive to transmit even small amounts of digital data. The Kavouras dialup systems widely used in the 1980s assigned intensities as 3-bit numbers, which allowed 2^3 or 8 levels of intensity. The WSR-88D improved this slightly by using 4-bit numbers, which provides for 16 levels. With data having become extremely cheap in the new millenium, the new super resolution and polarimetric products express quantities as 8-bit numbers, providing 256 intensity levels and showing

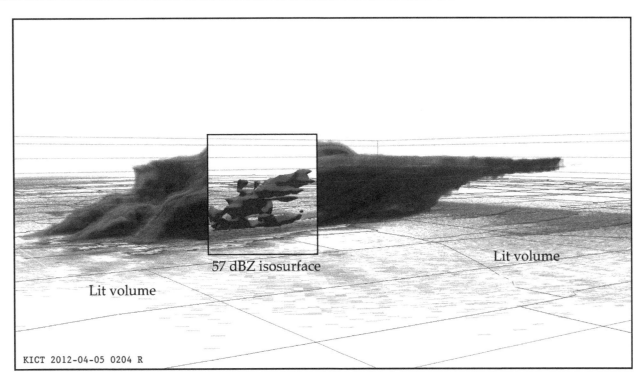

Figure 2-11. 3-D visualization of radar data using GRLevel2 Analyst (GR2AE). The image is constructed using the "lit volume" mode, and an inset of the "isosurface mode" has been overlaid manually at the center. In general, the isosurface mode is the most useful for applying 3-D interpretation techniques. In either case, forecasters must be careful not to assume that the echo shape shown here corresponds to the cumulonimbus cloud. The radar antenna detects rain, hail, and ice crystals, while water droplets which make up cloud material and cumulonimbus towers will be mostly invisible on this imagery.

noticeably better color depth on graphics images. This is considered to be good enough for most display and end-user processing purposes. Data levels are important because they express not linear or angular resolution of a product, nor time resolution, but rather the resolution of the data values themselves.

2.6.6. SMOOTHING. Smoothing is a mathematical technique for removing the "jaggies" in radar imagery by averaging neighboring points together to present a more aesthetically pleasing image. It is frequently done when presenting radar imagery on broadcast television or for the general public, and it can be accomplished instantly by many radar display programs. It was highly useful during the 1990s when radar data was coarse and only used 3-bit levels. Now with the shift to 8-bit levels and the widespread use of Level II super resolution, the need to smooth images has arguably diminished. Regardless of preference the technique is not recommended for critical and small-scale interpretation work since there is inherent information loss in the averaging process. Under no circumstances should it be done on small-scale velocity images because here, gradients and small-scale detail in the fields are especially critical.

2.6.7. MOSAIC. A mosaic is the term for an image formed from multiple radars, usually to create a regional or national radar product. In the case of an array, the parameters down to the individual gates are carefully blended and can be used as a basis for storm-scale interpretation. Mosaics, however, are primarily built by stacking images (typically composite reflectivity) and are best for getting a

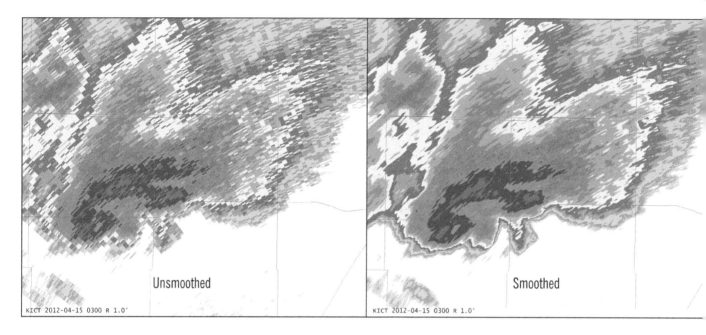

Figure 2-12. Smoothing of radar data, as seen with a reflectivity display of a tornadic supercell in Kansas. Though smoothing is often desired when using coarse data, it can actually complicate interpretation since the forecaster is no longer looking at the actual data but a mathematical approximation of it. Smoothing becomes less of a problem with high-resolution data, such as the Level 2 Super Resolution data seen here. The mathematical model is much more likely to closely approximate the real field with high-resolution data. With critical fields such as velocity, the use of smoothing is generally discouraged. In many cases, simply modifying the color palette may provide a much better image.

general overview of conditions across a broad area. They should not be used in storm-scale forecasting.

2.6.8. VOLUMETRIC FEATURES. Volumetric features are significant features that are found on nearly all radials. They provide meaningful information about the large-scale environment. Examples of volumetric features are bright bands (a ring circling the radar which represents the melting level), large-scale velocity signatures (revealing the upper-level winds throughout the volume), and synoptic-scale boundaries. To make a proper inference about a volumetric feature, some concepts the forecaster must be aware of include which VCP the radar is operating in and the conical nature of each tilt.

2.6.9. LOCALIZED FEATURES. Localized features are features which are confined to a small area or a small set of radials. They include thunderstorm cells, couplets (small scale wind circulations), and mesoscale boundaries. When investigating localized features, forecasters must consider which direction the radar antenna is relative to the feature and how far away it is, which VCP the radar is operating in, the conical nature of each tilt, and whether nearby radars might provide a better or different view of the feature.

2.7. WSR-88D

The full range of United States radar network products are exceptionally easy to access as the government has an open access policy. The main question comes down to which tools to use to visualize the radar.

2.7.1. RADAR DATA ACQUISITION (RDA) UNIT. The WSR-88D radome and tower immediately come to mind as the most visible parts of the *radar data acquisition* (RDA) unit. This subsystem consists all of the hardware and software that handles the electromagnetic energy, including the transmitter, receiver, and the antenna.

The radar transmitter is housed in a building near the radome and consists of a klystron, a large vacuum tube the size of a refrigerator which produces enormous amounts of tuned electromagnetic energy. This energy travels through a waveguide to the tower and enders a feedhorn, a device mounted on an arm positioned in front of the radar dish. This feedhorn radiates the energy onto the dish itself, which acting as a parabolic reflector transmits the energy outward into the atmosphere as a focused beam.

This antenna weighs 2600 pounds and is 28 feet in diameter, giving the RDA tremendous sensitivity. It is mounted on a sturdy aluminum and cast iron *pedestal* that contains all the motors and bearings needed to position the antenna with great accuracy. All of this is protected by a fiberglass radome 39 ft in diameter, which is transparent to the radar beam and is coated with a hydrophobic paint that repels rain and ice buildup. The radome also has lightning arresting spikes that safely conduct electrical charges down to the ground. The radome and all the equipment inside is is placed on a steel tower that may be up to 98 feet in height, depending on needs at each location.

During the listening window, the parabolic dish is used as a giant listening antenna that receives the backscattered radio energy. This energy is routed to a radio receiver near the tower, which turns the analog energy into an analog signal. This analog signal, paired with calibration and synchronization data and information on the antenna position, is referred to as the Level I datastream and is only accessed by engineering staff.

This analog information is fed to a signal processor which converts this data into digital form. It extracts the components of reflectivity (power), velocity (phase shift), and spectrum width (diversity of the phase shift). With the new polarimetric radar upgrade, three other products are transmitted: differential reflectivity, correlation coefficient, and differential phase. The signal processor also takes on the job of filtering ground clutter, range unfolding, and data thresholding in order to remove anomalous data before it is released to the RPG. The RDA produces the radar's set of base products, which are transmitted to the RPG and out to the world as the Level II datastream. This data was unavailable to forecasters when NEXRAD was first fielded, but by 2006 it was widely available over the Internet and is now a staple of forecast use since it offers the highest resolution available.

2.7.2. RADAR PRODUCT GENERATOR (RPG) SYSTEM. The RDA feeds the base products to the *radar product generator* (RPG) system. It is essentially a dedicated computer that uses the base products to generate derived products like storm-relative velocity, storm tracking, velocity wind profile, hail detection, precipitation totals, vertically integrated liquid, and many other types. All of this is transmitted to the world as the Level III datastream, which is used extensively by forecasters. The RPG also had dial-in "narrowband" capabilities that offered products to authorized users. Due to the reduced cost and dependability of broadband data links, data is now distributed almost entirely through weather telecommunications networks and the Internet.

2.7.3. PRINCIPAL USER PROCESSOR (PUP) SYSTEM. The PUP was the fancy name for the graphics workstation at National Weather Service and Air Force weather offices that displayed the WSR-88D data. It was considered to be the third and final part of the WSR-88D system, originally consisting of a pair of CRT monitors, a large graphics tablet, and a rack-mounted mainframe computer that controlled the workstation. The PUP was phased out in the late 1990s due to the arrival of the AWIPS system, a graphics computer system used by the National Weather Service which has the ability to interact directly with the RDA and RPG and display graphics products. This and the widespread availability of many third-party computer applications to display all of the WSR-88D products has made the PUP concept obsolete. Additionally, the old PUP only used Level III data, so in some respects the solutions available to hobbyists on PCs today are superior than the PUPs used by NWS forecasters in the 1990s.

2.7.4. CONTROL TERMINAL. The legacy WSR-88D was controlled at a unit control position (UCP) terminal, located at the closest federal weather station rather than at the radar site. This was replaced in the early 2000s with a Human-Computer Interface (HCI) workstation. This made it much easier to set environmental and adaptable parameters and to import observed data from the AWIPS data network.

2.7.5. DATA STREAMS. Forecasters must fully understand the three primary data streams used in WSR-88D operations. The Level I products consists of the raw, pre-processed analog output from the radar and is only available to engineering staff. The Level II products consist of the RDA's base products: reflectivity, velocity, and spectrum width. With the introduction of dual-polarization radar, it also adds several polarimetric base products. The Level III products encapsulate the products produced by the RPG, offering many more products beyond the base products but without the high-resolution available in Level II.

Controlling the radar

Unlike first and second generation weather radars, forecasters and radar operators are not able to manually control the WSR-88D's antenna. It is controlled completely by the Radar Data Acquisition system. The only individuals who can take direct control of the antenna are engineers onsite who are performing maintenance work.

Obsolescence in the WSR-88D

Originally the RPG used a Concurrent 3280 minicomputer, which was state of the art when it was released in 1985 and was priced at about $150,000 after volume discounts. It offered about the same horsepower as Intel 486 computers that hit the market in in 1993. The RPG worked with only 8 MB of memory and a 140 MB hard drive. Obviously the capabilities of the original RPG sometimes reached their limits in fast-breaking weather situations. A sophisticated load shedding program ran in the background which was designed to remove products from the queue that would likely not be needed. With the rapid increases in affordable computing power since then, the RPG has been significantly upgraded and most load shedding problems have been eliminated. As far as the Concurrent 3280, circuit boards have appeared occ

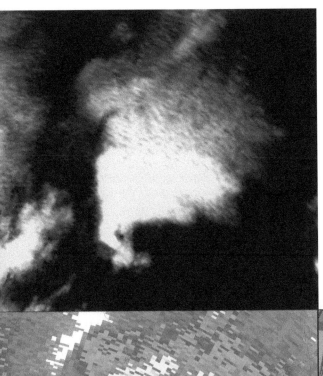

Figure 2-13. **WSR-88D Level I (top left), Level II (lower left) and Level III (lower right) data are compared** in this illustration. The Level I consists of analog output from the RDA and is never used by forecasters, nor is there easy access to this data. The Level II data, which in this case benefits from the 2008 Super Resolution upgrade, features an angular resolution of 0.5° and a radial resolution of 0.25 km. It offers the sharpest look at features within the storm, but at the cost of higher bandwidth. The Level III data, by contrast, has an angular resolution of 1° and a radial resolution of 1 km. The advantage of the Level III datastream is that it is associated with a large variety of derived products and the bandwidth requirements are lower. Forecasters will choose to use only L2, only L3, or both, depending on which sources are easily available, the workload, and personal preference. The subject of the image is the Irving, Texas supercell of 3 April 2012.

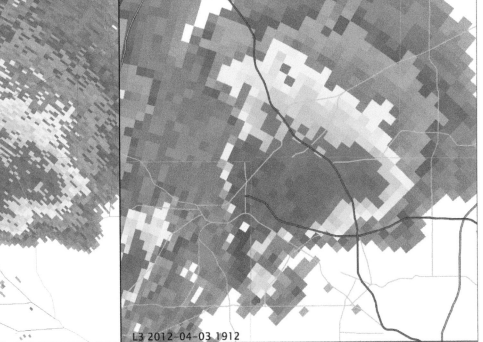

Data size comparisons

Here we list the size of Level II and Level III data. These are compressed data sizes. The "modern day" examples provide a useful estimate of Internet bandwidth requirements when monitoring a weather situation (up to 174 MB and 47 MB for an hour of Level II and Level III data, respectively).

Early WSR-88D / Quiet day
14 March 1996 (KJAX)
Level II - 131.7 MB
Level III - 23.7 MB

Early WSR-88D / Storm day
5 May 1995 (KFWS)
Level II - 254.0 MB
Level III - 26.3 MB

Modern WSR-88D / Quiet day
16 November 2012 (KFSD)
Level II — 225.28 MB
Level III — 103.42 MB

Modern WSR-88D / Storm day
30 October 2012 (KLWX)
Level II — 4194.3 MB
Level III — 1139.7 MB

Though it might seem Level II is a popular choice for experienced users, the velocity products from Level II are not dealiased, and storm-relative velocity, which is important to storm-scale analysis work, is not offered. This is a setback, but at the same time it allows the user's software to accomplish these tasks and create these products with much more flexibility. For example, the Level III storm-relative velocity products use a specific wind vector which is set at the RPG, but with a Level II viewer that can produce storm-relative velocity, such as GRLevel2, the user can vary the storm motion vector to improve the signatures of small circulations and other features.

Even though a typical WSR-88D radar produces many gigabytes of data per day, United States weather agencies have a mass storage system for the archival of WSR-88D radar, with free access to the public. This system, comprised of over 1300 terabytes of data, can be found at <has.ncdc.noaa.gov>. The Level 2 and Level 3 data can be plotted using NCDC Java-based viewers or off-the-shelf software like GRLevelX.

2.7.6. VOLUME COVERAGE PATTERN. The WSR-88D radar is engineered to operate in specific modes, each with its own set of PRF, elevations, and scan rate. These are known as volume coverage patterns (VCPs). Initially the radar had two "storm modes" (11 and 21) and two "clear air modes" (31 and 32), however after several years in service a number of new VCPs were introduced to enhance the capabilities of the radar. The VCPs are fully listed in the appendix.

2.7.7. ENHANCEMENTS. In 2008 the National Weather Service began upgrading its radars to disseminate a "super resolution" data format, in which all of the base products are offered with 0.5° of radial resolution rather than 1°, and with gate resolution of 0.25 km instead of 1 km. This super-resolution capability is only disseminated in the Level II datastream.

In 2011-2013 the National Weather Service upgraded its WSR-88D network to convert the network from conventional radars to polarimetric radars. Since the capability to transmit and receive the horizontally-polarized signal is unchanged, there is absolutely no effect on any of the standard conventional products. Rather this upgrade simply adds new capabilities and products to the existing ones.

2.8. TDWR

In response to a series of fatal wind shear incidents, including a 1982 Delta 727 crash in New Orleans and a 1985 Delta L-1011 crash at Dallas-Fort Worth Airport, the Federal Aviation Administration quickly began installing wind shear sensors at major airports and began exploring proactive rather than reactive solutions. The TDWR was largely designed by MIT's Lincoln Laboratory in

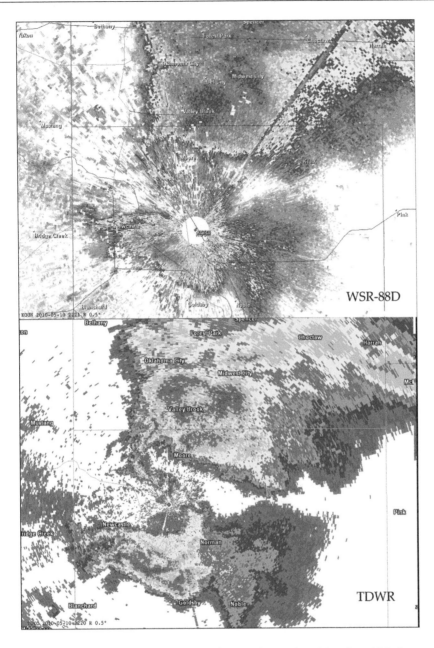

Figure 2-14. Comparison of WSR-88D (top) and TDWR (bottom) radars for roughly the same instant and transmitting from a location only 4 miles apart. It can be seen that the output is very similar, though the C-band TDWR shows attenuation at ranges that are deep into the precipitation (top of images).

collaboration with NCAR and NSSL during the mid-1980s. In 1988 the FAA locked down the design in 1988 and awarded a production contract to Raytheon. The first TDWR was installed at Memphis in 1992. Although it has clearly been around for over 20 years, it is a fairly new system to many meteorologists since the TDWR output was not added to NOAA datastreams until 2009.

As of 2012, 48 units were in service in the United States, most of them in major cities. Since most WSR-88D units have been intentionally sited at least 30 miles away from large cities in order to keep critical storm analysis from being degraded by clutter and the cone of silence, the TDWR a very useful overlap to the weather radar network, and all forecasters should be aware of how to access TDWR products if there is a site near their area.

A quick look at VCPs
VCP 11 - 14 tilts in 5 min
VCP 12 - 14 tilts in 4.5 min
VCP 21 - 9 tilts in 6 min
VCP 121 - 9 tilts in 5.75 min
VCP 31 - 5 tilts in 10 min
VCP 32 - 5 tilts in 10 min
VCP 211 - 14 tilts in 5 min *
VCP 212 - 14 tilts in 4.5 min **
VCP 221 - 9 tilts in 6 min *

* - Uses SZ-2 range unfolding algorithm on the lowest 2 tilts
** - Uses SZ-2 range unfolding algorithm on the lowest 3 tilts

2.8.1. DESIGN. The TDWR operates at 5.6 GHz (5 cm, or C-band) at a power of 250 kW, a third that of the WSR-88D. It has a beamwidth of 0.55° in comparison to the WSR-88D's 0.95° figure, and has a range of 248 nm (48 nm for the velocity channel). It also has a maximum antenna tilt of 60°, as compared to the WSR-88D's 19.5°, so there is capability to overcome the cone of silence by using extremely high tilts, though many tilts must be skipped for practical purposes and this makes elevation gaps a problem. The TDWR has a range resolution of 0.15 km compared to the WSR-88D's 0.25 km super resolution.

Since the TDWR is a C-band radar, it is more vulnerable to attenuation. Therefore forecasters need to be on guard when using the radar with mesoscale convective systems or when one cell is being shadowed behind another along a given TDWR radial.

2.8.2. SCAN STRATEGY. The TDWR, like the WSR-88D, has a number of volume coverage patterns. There are 17 tilts when in clear air mode and 23 tilts when in precipitation mode. A complete volume scan takes 6 minutes.

2.8.3. AVAILABILITY. The Level III TDWR data is distributed without restriction to the general public, and can be plotted with programs like GRLevel3. The higher-resolution TDWR Level II data is not distributed outside the FAA and NWS, presumably due to the additional distribution costs.

REVIEW QUESTIONS

1. Given a wavelength of 3000 MHz, convert the frequency to wavelength.

2. If a pulse repetition frequency is 1000 Hz, how far will the pulse travel before the radar emits another pulse?

3. What does increasing pulse repetition frequency (PRF) do to the maximum unambiguous range, and why?

4 Name the two key types of scattering, and which precipitation forms they predominate in? Using only a few words, in which direction is the radar energy reflected?

5. How does a bin change in terms of size and altitude as it goes to farther ranges along a radial?

6. Why is it impossible to detect a tornadic circulation at a range of 300 miles?

7. The two key types of operational weather radars are C-band and S-band radars. Identify the wavelength, frequency, and general purpose of these two radars.

8. Name the two major components of a WSR-88D radar site.. Describe which major products originate from the WSR-88D RDA.

9. List the three conventional base products of the WSR-88D. Bonus: also list the three dual polarization products.

10. What is one way that the TDWR differs significantly from the WSR-88D, aside from its use in the aviation industry?

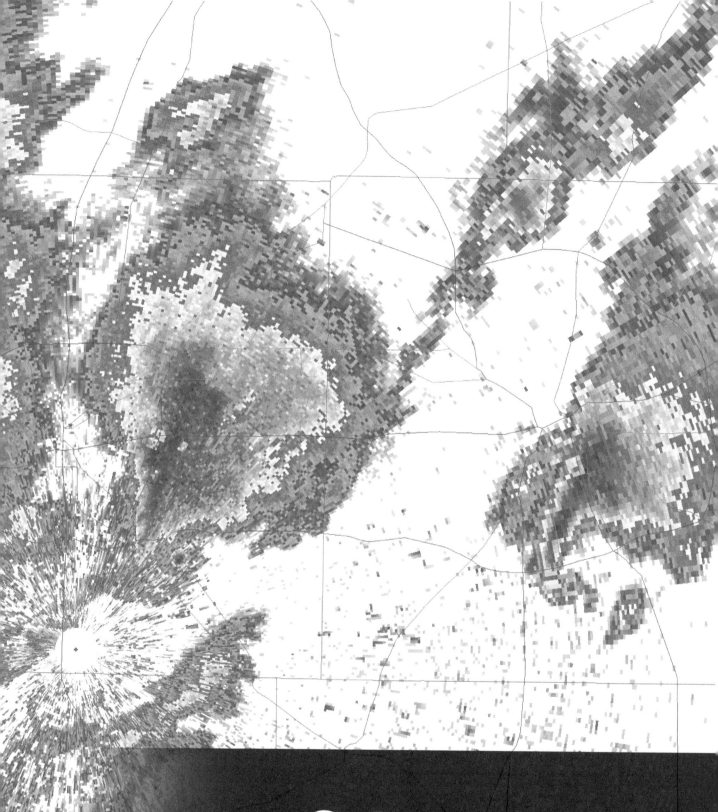

3 Reflectivity

Reflectivity is the most universally recognized product of radar technology. Since it has been used in meteorology for 70 years, it has been extensively studied. It makes up the backbone of storm detection and is available wherever a radar unit exists. For this reason it is referred to as conventional radar. Even for those in the United States who enjoy a weather agency that is generous with the dissemination of advanced products like velocity and polarimetric data, conventional radar interpretation techniques form the basis of nearly all operational storm analysis methods. Such methods were widely used from the 1960s to the 1980s and refined since then. So conventional interpretation techniques deserve a section all their own.

Given the rapid advances in velocity and polarimetry and the withering away of reflectivity interpretation topics in journals, this does not mean that the use of reflectivity is becoming marginalized. The advanced products are largely useless without reflectivity, and it still remains a solid starting point for all weather forecasting analysis and interpretation. When things get confusing, a careful review of reflectivity products often puts things back in place. Forecasters must remain solidly acquainted with conventional interpretation, understand its limitations, and comprehend what effects might spill over into velocity and polarimetric products. Furthermore, reflectivity still remains a staple of radar meteorology outside of the United States, where advanced parameters are largely unavailable or limited to forecast offices, universities, and corporate weather offices.

3.1. Problems affecting reflectivity data

Though reflectivity is a rather simple quantity, there are a few issues which may affect reflectivity products, aside from the standard radar propagation anomalies that are listed in the previous chapter. Here we will focus on some of the more specific problems.

3.1.1. CLUTTER. A nonmeteorological target is usually referred to by the forecaster as ground clutter or anomalous propagation. Although this data is usually seen as an annoyance, airborne clutter (insects, dust, and pollen) does provide a layer of scattering for velocity and polarimetric products that may not otherwise be present.

The largest source of clutter occurs close to the radar site as the radar beam and its sidelobes interact with objects on the ground, particularly at the lowest tilts. As a result, the radar data is usually contaminated within several miles at low elevations. Large buildings and highways may be seen and recognized on the imagery.

The WSR-88D's signal processor uses a predefined clutter suppression map to filter out this type of ground clutter. In addition to this, the radar operator can define additional clutter suppression regions. Due to slight differences in

Title image
Multiple supercells rake the Dallas-Fort Worth metroplex on 3 April 2012 at 1:26 pm CDT, as seen by the FWS 0.5° base reflectivity image. The storm on the left was a supercell containing a tornado producing EF2 damage in Arlington. The storm on the right, also a supercell, was moving through extreme south Dallas and also contained a tornado producing EF2 damage. The tornado is at the tips of each of the "hook" appendages on the south side. The Arlington tornado shows a pronounced "debris ball". This image was produced with Level II super-resolution imagery.

refraction of the atmosphere, the ground clutter may be weaker or stronger from day to day, introducing variability, so it cannot be completely removed.

Outside of the ground clutter area, nonmeteorological signatures may be produced by smoke plumes from wildfires and ash clouds from volcanoes. These can be very useful to emergency operation centers. The appearance of birds and insects, particularly during the dusk hours, have been well documented. The emergence of bats from their habitats are also a unique feature to the Austin-San Antonio (KEWX) and Del Rio, Texas (KDFX) radars during the spring and early summer, when they fly up to a thousand feet high or more to eat moths riding north from Rio Grande breeding habitats to crop fields further north in Texas. Aircraft may reflect radar energy, and will produce very isolated blips that disappear with the next volume scan.

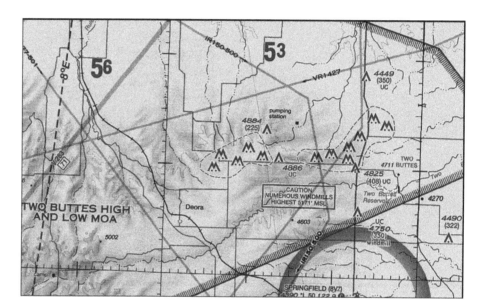

Figure 3-1a. Wind farms on FAA aeronautical charts. These charts are available at skyvector.com or <http://aeronav.faa.gov> under "Sectional Charts". These charts are updated twice a year, so they are a timely source of information on confirming wind farms and other obstructions. Here, the Colorado Green Wind Power Project can be seen, showing towers rising to 408 ft AGL. It can also be seen that the area is in a "MOA" (military operating area); these and restricted or warning areas (airspace marked with an R or W preceding a number) may be a source of chaff.

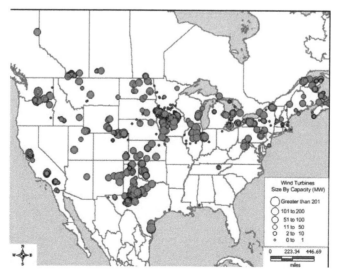

Figure 3-1b. Wind farms in North America as of 2011." Though clutter suppression maps have largely eliminated them from weather radars, they may occasionally be visible. New wind farms under construction may also appear on radars. *(U.S. Energy Information Administration)*

An undesirable signature is produced by chaff. Strips of metallic material known as chaff are often released by military aircraft operating in busy training areas. Two notable areas in the United States that are affected by chaff are the Luke Range southwest of Phoenix, Arizona and the Nevada Test and Training Range northwest of Las Vegas. Chaff is designed to produce a "cloud" of high backscattered energy to enemy radars, hiding planes flying within it. As a side effect, it causes enormous blooms of false echoes on weather radars. This not only puzzles the forecaster but can contaminate a wide range of products from precipitation estimates to climatological studies which use archived radar data. Meteorologists should be aware of military operating areas around their forecast areas and crosscheck unusual patterns with other meteorological tools.

One additional backscatterer has risen to prominence in the past decade: wind turbine clutter (WTC). This is produced by dozens, and in some cases hundreds of turbines at wind farms, which have begun dotting the Great Plains, Great Lakes region, and other areas in the United States at an amazing rate over the past ten years. Not only do the turbines backscatter reflectivity, but they also produce signatures for velocity and spectrum width products. They also cannot be removed through normal clutter suppression techniques.

3.1.2. REFRACTION. The pressure, temperature, and moisture in the atmosphere drops sharply with height, yielding a specific density profile. The speed of light is dependent on the density through which it moves, and any change in velocity will cause a change in direction, called refraction. The density does drop rapidly with height and this alone allows the radar beam to refract somewhat over the visible horizon. However, an unusual rate of cooling or warming with height may significantly alter the refraction and cause the radar beam to point more upward or downward than expected.

For example, if the atmosphere cools with height at a weaker rate or warms with height (an inversion), this will bend the radar beam down to the ground and result in *superrefraction*. Backscattered energy will be at a lower elevation than expected, and the radar beam may even reach the ground at a long distance. If the beam travels a long way, matching the curve of the earth somewhat, it is said to have undergone ducting. Ducted radio waves may travel thousands of kilometers before reaching the ground or departing the atmosphere. This the phenomenon has been used at night for many decades as a way of sending shortwave radio transmissions over very long distances.

Likewise, if the atmosphere cools sharply with height, such as in a highly unstable air mass, the radar beam will tilt higher than expected and result in *subrefraction*. The overall effect is that this may cause the radar analyst to misinterpret the height of storm tops and other phenomena.

On a day-to-day basis, the refraction artifact that a forecaster is most likely to encounter is superrefraction in the clear air behind squall lines and other convective weather systems. When refraction anomalies are suspected, the forecaster can choose a different tilt, if available. Derived products may contain

---

Refraction problems

Superrefraction
This tilts the beam to a lower elevation. It occurs where within a layer, there is:
* Strong warming with height (e.g. an inversion)
* Strong moisture decrease with height
* Warm air moving over a cold surface
* Outflow from thunderstorms

Subrefraction
This tilts the beam to a higher elevation. It occurs where within a layer, there is an unusually strong:
* Steep lapse rate
* Strong moisture increase with height

errors, particularly those which rely upon accurate measures of altitude, such as the hail detection algorithm. Radials which are not suspected to lie above an area with an unusual temperature profile will probably not be affected by refraction anomalies.

3.1.3. RANGE FOLDING. The problem of range folding was outlined in detail in the Fundamentals chapter. Range-folded echoes, if they appear, will show in unexpected areas and will take on a elongated appearance. For the most part, modern third generation radars can unfold reflectivity data. The most popular technique is called interlacing, where the radar alternates two wavetrains that have different PRFs. This will cause second-trip echoes to move, while the primary echoes stay in the same place, and this can be easily detected and corrected by the radar's algorithms. Velocity products, however, depend on one PRF and as a result are often heavily impacted by range folding.

As described earlier, a second trip echo consists of backscattered energy from a meteorological target outside of the R_{MAX} radius of the radar. It may occur if the radar is using too high of a pulse repetition frequency. Second trip echoes usually look like elongated streaks of echoes in an area where no weather is known to be occurring.

3.1.4. WET RADOME. Though most radomes are designed to be hydrophobic and repel water, a coating of water may still cover the surface in heavy rain or freezing rain situations. This will reduce the power transmitted and received, in turn causing a reduction in the reflectivity. While this will have little effect on most products, the precipitation products are sensitive to this loss of energy, and it may cause rainfall totals to be underestimated.

3.1.5. SUN STROBES. When the radar dish is pointing into the rising or setting sun, a spike may be seen extending in this direction. This is known as a sun strobe. It is caused by the reception of electromagnetic energy from the sun, and is usually gone after 1 or 2 scans. Its appearance at a specific azimuth and elevation can be predicted with such accuracy that it was used to check the antenna calibration of older second generation radars.

3.2. Precipitation

The current generation of radars have tremendous ability to accurately measure reflectivity of pulse volumes within the cloud at small scales. However this comes with several assumptions.

3.2.1. ASSUMPTIONS. First, there are the standard beam behavior assumptions; refraction, beam blockage, and so forth are expected to be insignificant. Radar interpretation assumes that all backscatter is from Rayleigh scattering, the precipitation falls vertically in its entirety, and precipitation particles are homogenous and fills the pulse volume evenly.

This assumption breaks down in the presence of large particles, as was discussed in the previous chapter where Mie scattering occurs if particles are sufficiently large. Precipitation estimates are largely based on the reflectivity-rainfall (Z-R) relationship which involves Rayleigh scattering. Polarimetric radar, which is discussed in an upcoming chapter, offers enormous potential for differentiating precipitation types and improving precipitation estimates.

Our assumption about precipitation reaching the surface breaks down if strong winds underneath the beam are occurring, which can advect precipitation as it falls, dispersing it away from areas where rainfall is expected and introducing it in locations where none is expected.

Evaporation beneath the beam is another problem. For example, in a very dry air mass, virga will predominate and little precipitation will actually reach the surface.

Coalescence below the beam may affect precipitation estimates in tropical air masses. Such an air mass has a very large number of small droplets. As a result, there may be higher intensities below the beam that are not being detected, and underestimation of rainfall amounts results.

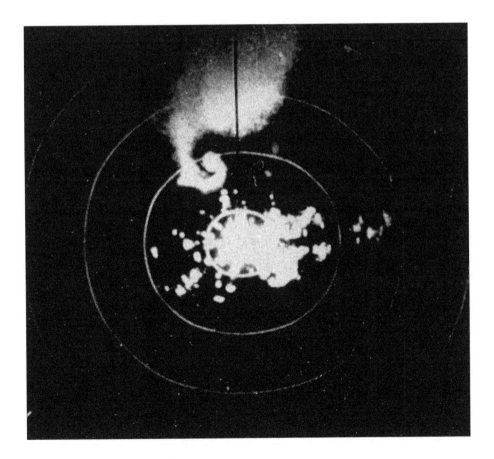

Figure 3-2. Hook echo on 19 May 1960 at the Topeka, Kansas WSR-3 radar site. Hook echoes had been observed for many years, but this is a particularly distinctive example. At the time, a tornado was moving through Meriden, Kansas. *(National Weather Service)*

3.2.2. BRIGHT BAND. The so-called "bright band" is often seen in stratiform precipitation events during the cool season, showing as a ring around the radar site. It is especially visible at higher tilts. It must be remembered that all of the tilts are conical, so the bright band simply represents where the beam is ascending through an altitude where enhanced reflectivity by wet snowflakes are present. The forecaster must remember that range is proportional to altitude. The outer edge of the bright band is at the melting level itself, at the altitude where dry snowflakes are transitioning to wet snowflakes. The inner edge of the bright band shows the altitude where wet snowflakes are transitioning to rain.

If the display software shows altitude at the cursor location, or the range can be obtained and converted using a nomogram, the exact numerical altitudes can be determined. Also by keeping track of trends of the altitude of the bright band and comparing them to other radars, the forecaster can get an idea of its slope and get information on when rain might begin changing to wet snow at a given location.

3.3. Thunderstorms

The body of knowledge concerning convective weather and thunderstorms, even at the operational level, is vast and is well beyond the scope of this book. Forecasters do need to be well-grounded in this area in order to make the best use of radar products. Since this book centers on radar meteorology, only a very basic framework of storm forecasting knowledge will be presented. Readers who are weak in knowledge in this area are strongly urged to study books, journals, and other resources relating to severe thunderstorm forecasting for a much more complete understanding.

3.3.1. CONCEPTUAL MODEL. The thunderstorm is made up of two basic processes: an updraft and a downdraft. The updraft is fed by inflow, is manifested by the appearance of a cumulonimbus cloud, and consists mostly of cloud droplets which are not detectable by normal network radars. The downdraft is produced by condensation and droplet growth originating from the updraft, is manifested by a region containing precipitation or evaporatively cooled air, and produces outflow as it reaches the surface.

3.3.2. SEVERE WEATHER SIGNATURES. Many of the techniques used in *operational* severe weather forecasting were initially pioneered by National Severe Storms Forecast Center technician Les Lemon in the mid-1970s, drawing upon experience and the existing research knowledge at the time.

One of the first differentiators between a severe and non-severe event often involves whether a cell is isolated or not. Given the same environment, an isolated cell has less competition for moisture (fuel) and is less likely to encounter cell interactions that might interfere with processes leading up to the production of large hail, high wind, and tornadoes. Multicell situations can produce severe weather, of course, and this will be covered in the next section, but a strong and isolated cell will have the greatest potential for severe wether.

Isolated storms show relatively distinct identities and appear to be separate from other nearby storms. In nonsevere environments these usually consist of weak multicell storms and the rare weak unicell storm. In severe environments it may consist of a strong multicell line or a supercell storm.

At the individual cell level, the appearance of *strong intensities* and formation of strong *intensity gradients*, particularly adjacent to the updraft, are very important indicators of a transition to a long-lived cell mode or, further, to severe modes. The shift of strongest intensity from the cell center to the cell edge is often yet another indicator. The growth of high reflectivities to high altitudes above the ground has been identified as an indicator of storm severity. One basic technique outlined by Lemon is the existence of 50 dBZ at 27,000 ft AGL or higher, or a 45 dBZ return anywhere between 16 and 39 thousand feet AGL. This rule was developed mainly for "classic" springtime storms on the Great Plains.

Severe storms normally exhibit a significant amount of *tilt*, which is an axis connecting the highest intensities at the surface to the highest intensities in the upper portion of the storm. In severe storms the tilt (looking from bottom to top) tends to point to the direction of the storm inflow. Note that this is not equivalent to visual tilt of the cumulonimbus cloud, nor is it influenced by environmental shear; rather the tilt is simply an axis that forms due to changes in the storm's structure.

Isolated storms may take on a triangular shape, with the triangle appearing to point upwind, and have the appearance of a flying eagle. They show very strong gradients on the upwind side. This is the typical signature of a supercell thunderstorm in the absence of a hook echo (described below).

The kidney bean pattern is associated with isolated HP supercell storms. The concavity on the forward side of the storm contains the updraft and wall cloud, flanked on three sides by shafts of hail and precipitation, and is often referred to by chasers as the "bear's cage".

3.3.3. UPDRAFT STRUCTURES. In addition to readily discernable signatures, severe storms may also show clearly defined updraft structures. The simplest type of severe structure occurs due to the acceleration of water and ice production above the updraft area and propagation of the cell into the inflow (usually toward the right in the northern hemisphere, assuming we are looking downstream with the mean tropospheric flow). This produces a pattern known as overhang, where significant reflectivities exist in the middle and upper parts of the storm above the updraft zone. The area of diminished reflectivity immediately beneath the overhang is known as the weak echo region (WER), and is a region where the updraft saturates and is rising so quickly that the water

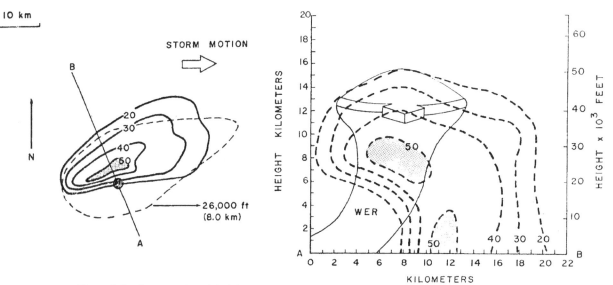

Figure 3-3a. Conceptual model of the weak echo region (WER), as developed and depicted by Les Lemon (1977) in NOAA Tech Memo NWS NSSFC-1. On the left is a horizontal low-level depiction of the storm with units of reflectivity (dBZ) drawn in solid isopleths. A thin dashed outline indicates where the upper-level "overhang" is located, and the line A to B is diagrammed as a cross section, above right. The weak echo region is that level below the overhang. Overall this pattern is indicative of a severe storm, and is normally associated with a severe multicell or supercell storm.

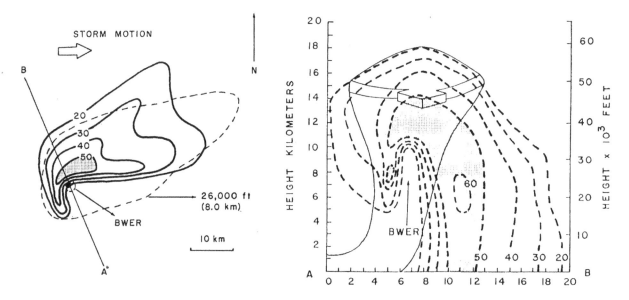

Figure 3-3b. Conceptual model of the weak echo region (BWER), same reference and layout as in figure "a" above. This mode represents an even stronger mode of cell intensity. Instead of a weak echo region, a bounded weak echo region (BWER) exists below the overhang. This model was explored as an operational predictor of tornadic storms, and it is certainly a strong indicator of a supercell.

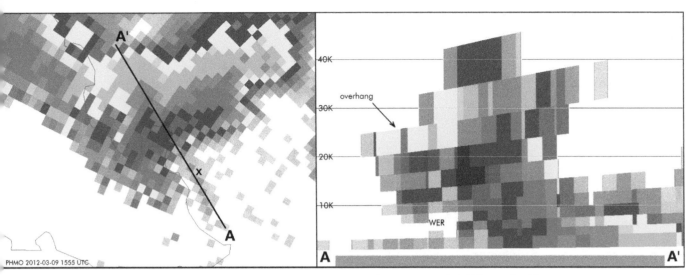

Figure 3-3c. The weak echo region as seen on a WSR-88D radar. The left image shows the 0.5° base reflectivity image, while the right image shows a vertical cross section along the line marked A-A'. The weak echo region consists of the broad concave region on the southeast side of the storm. A peak reflectivity of 70 dBZ was detected at a height of 19,000 ft above the "X" mark on the horizontal image at left, where the 0.5° reflectivity only showed 11 dBZ. This is the 9 March 2012 supercell near Kaneohe, Hawaii, which produced a spiked hailstone measuring 4.25 inches long and 2 inches wide, and was confirmed by NWS Honolulu.

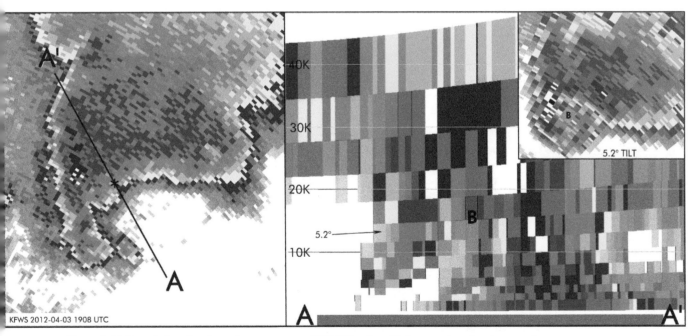

Figure 3-3d. Bounded weak echo region as seen on a WSR-88D radar, with the same layout as the preceding set. The BWER is marked by a bold "B" in the cross section, where a 30 dBZ core is surrounded by very intense reflectivity. The corresponding location is shown with an asterisk in the left image along the cross section line. An inset showing the 5.2° tilt is provided in the upper right corner, showing the "donut" of high intensity surrounding the BWER core. This was a tornadic supercell moving through northern Irving, Texas on 3 April 2012.

> **The hook echo in the 1950s**
>
> "To catch the formation of significant 'hooks' on radar echoes the operator must observe the scopes continuously during a severe storm situation; otherwise he may miss the hook completely or, if he observes it, be unable to tell whether the appendage is a true hook. If further experience with radar echoes shows that, as in this case, hooks appear well in advance of tornadoes, it may be possible in favorable circumstances to issue advance warnings on the basis of the PPI scope presentations before the tornado drops from the clouds."
>
> - Alexander Sadowski, 1958
> *"Radar Observations of the El Dorado, Kansas Tornado, June 10, 1958"*

droplet growth process is not completed until the parcels are at the overhang level.

If the updraft strengthens even further, the region of delayed precipitation formation actually punches upward into the overhang by as much as a few kilometers. This marks the transition of a WER into a bounded weak echo region (BWER). This appears on reflectivity as a small "hole" or small spot of diminished reflectivity embedded within the higher intensities of the overhang. The weak reflectivity opens in a downward direction, not upward.

Lemon's original technique specified that if the overhang extends more than 3.2 nm from the strong low-level reflectivity gradient, then this indicates a severe storm. Lemon's technique also emphasized that if the highest echo top was directly located over the low-level reflectivity gradient, then a severe thunderstorm is indicated. A common problem with this technique arises in storms containing large, wet hail, causing spikes due to three-body scattering and distorting reflectivity patterns at the top of the cloud.

3.3.4. HOOK ECHO. The fact that "hook echo" has come close to entering the English lexicon as a household word hints at its effectiveness as a tornado indicator. It was identified in the early 1950s and originally described as a pendant, with hook becoming the preferred word by the late 1950s. The hook tends to form on the southwest side of a supercell (northwest in the southern hemisphere) and appears as a cyclonically-curved finger or bulge measuring up to 5 to 10 km long. It is a mesocyclone signature, not a tornado signature, and is caused by the wrapping of precipitation into the mesocyclone. Any tornadoes, if they occur, tend to develop near the tip of the hook, with some tornadoes showing a debris ball. The available precipitation and strength, size, and depth modulate the appearance and structure of the hook echo. With LP (low-precipitation) supercells, no hook echo will typically be seen. The CL (classic) supercell is associated with the typical hook echo, while HP supercells contain a very dense, thick hook that gives the storm more of a "kidney bean" shape on radar.

3.3.5. DESCENDING REFLECTIVITY CORE (DRC). The descending reflectivity core was defined by Rasmussen et al (2006) and is characterized by a downward-pointing "finger" of relatively high reflectivity in the right rear quadrant of a right-moving supercell that lowers to the surface over a period of about 5 to 15 minutes. It is thought to form the hook echo. Finding a DRC requires keen mental visualization skills using multiple tilts, or better yet a 3-D display of reflectivity that paints "isosurfaces" of a specific reflectivity. This is a feature available in the GRLevel2 software program. There is no ideal value for DRCs, with possible isosurface choices ranging anywhere from 30 to 55 dBZ, and the ideal value for viewing a DRC can change even from scan to scan.

Though any area of high reflectivity suspended aloft is comprised of hail and rain and will eventually descend or "collapse" with time, the DRCs of interest

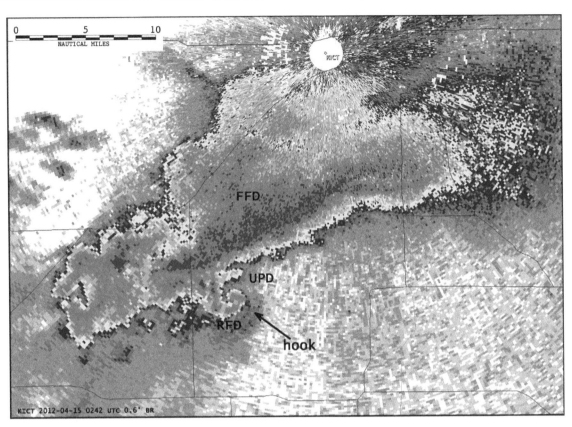

Figure 3-4. Hook echo associated with a half-mile wide tornado southwest of Wichita KS on the evening of 14 April 2012. The main parts of the supercell have been marked FFD (forward flank downdraft), RFD (rear flank downdraft), and UPD (updraft)..

are localized fingers of higher reflectivity close to the mesocyclone and which descend with successive scans. It is speculated that they might not only form the hook but may be associated in some way with tornadogenesis. Further research is still needed to explore this link and whether the pattern offers any definitive forecasting cues.

3.3.6. Vortex Hole (VH). The concept of a vortex hole in supercells was proposed by Lemon and Umscheid (2008). This is a column of weak reflectivity embedded within or near the overhang, surrounded by a "wall" of strong intensities. It was proposed that this vortex hole might be a cylindrical vortex in which water and hail is centrifuged away from the center. The vortex hole is differentiated from the BWER in that the latter is a feature extending underneath the overhang upward, while the vortex hole is embedded in the precipitation and may extend to the top of the storm.

3.3.7. Splitting Storms. Though weak storms occasionally seem to "split" and form new cells, eventually aggregating and forming multicell clusters, storms in severe environments may become very intense and even form supercell structures, before splitting with one half moving to the left of the mean environmental winds (typically to the north) and the other half moving

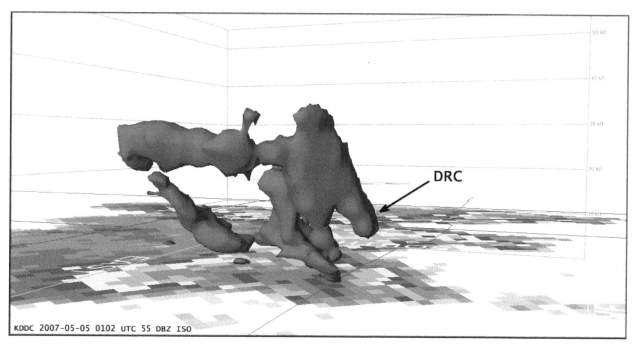

Figure 3-5. Descending reflectivity core (DRC) near Bucklin, KS on the evening of 4 May 2007 as viewed using the 55 dBZ isosurface. The view direction is to the south. This was about 20 miles southwest of Greensburg KS, which was destroyed about an hour later by a different supercell.

to the right of the mean winds (to the east). The phenomena of splitting storms has been documented since the 1950s. Early modeling quickly revealed that splitting modes were favored in environments where the hodograph shows a significant lack of curvature (i.e., the "straight line hodograph"). The simplest example of such an environment are where the winds are southwesterly at all levels, decreasing from strong winds aloft to calm at the surface. Such events may produce tornadoes, especially with the right mover, but most splitting cell episodes tend to result in many cell collisions and the eventual formation of an MCS.

3.3.8. Bow Echo. Another type of squall line deformation is where one part, or in some cases, the entire line accelerates forward, producing a forward-bowed appearance on radar. This is associated with increased efficiency in the storm's downdraft mechanisms on the back side.

One signature that is a hallmark of smaller, more intense bow echoes is the rear inflow notch. This is an area of diminished radar reflectivity, a "channel" of weak reflectivity, in the lowest 1-2 km of the storm on the back side of the MCS, often measuring only a few km in size. It signals the arrival of the rear inflow jet at the surface and may mark a transition from a linear, symmetric squall line into a bow echo. A rear inflow notch should be evaluated with velocity and polarimetric products, if available.

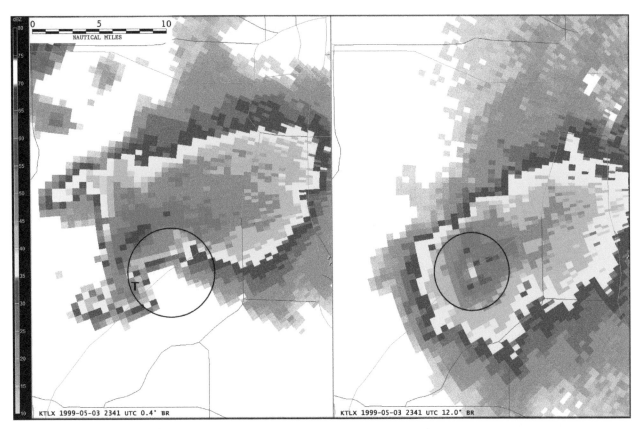

Figure 3-6. Vortex hole associated with the Bridge Creek OK F5 tornado on 3 May 1999. The vortex hole is at the center of the circle on the left panel (12.0°, at about 29,000 ft MSL) with the corresponding location shown on the 0.5° tilt at left. This indeed looks a lot like a BWER, but following the hole down to lower tilts brings it westward through the precipitation shaft to the tornado "T". The velocity product at 12.0° showed 100 kt of rotational shear between adjacent bins, clearly identifying this as a tornado in the upper parts of the storm. It's also a reminder of why commercial aircraft give storms like these a wide berth.

3.4. Squall lines

The unique structure of squall lines as a two-dimensional storm system was recognized in the 1920s and the general structure has not changed much since then. A squall line consists of a large rainy downdraft that produces outflow at the ground, coupled with a large cumulonimbus updraft fed by inflow. Normally the inflow area is found downwind from the outflow area, and the boundary between the two marks the squall line gust front.

Though severe weather in a purely two-dimensional squall line is limited to small hail and high wind, localized breaks from the two-dimensional structure may be associated with localized areas of damaging winds, small tornadoes, and possibly the appearance of LEWP and bow echo patterns as described below. Cells in a broken squall may intensify into severe multicells and supercells. Otherwise, the tail end of a squall line, known by storm chasers as "tail end charlie", is normally the preferred location for severe weather.

3.4.1. THE MESOSCALE CONVECTIVE SYSTEM (MCS). With the advent of mesoscale meteorology in the 1980s, many mesoscale systems were recognized as distinct systems of their own with pronounced effects on surface and upper level patterns. The squall line is one example of a large-scale mesoscale convective systems (MCS). Another example of an MCS is a hurricane.

3.4.2. LINE ECHO WAVE PATTERN (LEWP). Departures of an MCS from a symmetric, linear configuration were associated with severe weather and tornadoes. One of the first patterns to be described was the line echo wave pattern, where the long axis of the squall line takes on a wavelike or "S" appearance. The LEWP tends to be associated with squall lines that have broken up into one or more supercells, but where enough precipitation exists between the cells to give it a continuous appearance. The parts of the LEWP that form large concavities on the forward side, such as the top part of an "S" shape, correspond to strong sectors of inflow feeding a mesocyclone.

REVIEW QUESTIONS

1. Name at least two basic reflectivity signatures for severe thunderstorms.

2. Second trip echoes cause which type of radar anomaly?

3. If reflectivity shows a sudden bloom of high intensity echoes, including at higher tilts, and satellite data and surface observations show clear skies, what might this be?

4. What artifact appears on base reflectivity as a spike extending along a radial away from a hail core? Use the full name of the feature.

5. Storms with classic (CL) supercell characteristics tend to produce a textbook hook echo when they become tornadic. What is the corresponding signature in HP (high precipitation) supercells?

6. The clear area enclosed by a hook echo and the forward flank updraft of a supercell corresponds to what?

7. Overhang represents the tops of the storm appearing to "lean" into the inflow air. This overhang is a manifestation of what severe weather feature?

8. What is the difference between the echo-free region in the BWER and the echo-free region in the vortex hole (VH)?

9. The descending reflectivity core is associated with which part of a supercell?

10. What is the typical environmental wind profile in a situation favoring splitting storms?

4 Velocity

4 | Velocity Interpretation

Most scientifically-minded individuals understand the Doppler effect as the change in pitch that occurs when a loud object passes by at high speed, whether at an airport, racetrack, train track, or simply standing at an intersection waiting for the crosswalk light to turn green. The Doppler effect can be traced back to Austrian scientist Christian Doppler who proposed it almost 200 years ago to explain the colors of stars as being influenced by relative motion. The Doppler shift is easy to visualize: by picturing a train of waves rapidly radiating in all directions from a moving object, it makes sense that a stationary observer in front of the object would receive the waves at a faster rate. This principle is used to determine velocity characteristics from radar signals.

4.1. Velocity processing

Since the primary characteristic of a Doppler shift is the change in frequency, it is possible to simply measure this change to obtain velocity. If the backscattered energy has a smaller wavelength or higher frequency, this implies there is motion toward the radar, and if the energy has a larger wavelength or lower frequency, there is motion away from the radar, and the magnitude of the change is proportional to the velocity. Consider that we are sampling a tornado at a frequency of 2850 MHz. The part of the tornado where winds are blowing toward us at 50 m s^{-1} moves at about 10^{-7} of the speed of light, and the backscatter would have a frequency of 2850.00001 MHz. However this change is so small that it is difficult to measure accurately by direct measurement.

4.1.1. PULSE PAIR PROCESSING. The technique of *pulse-pair processing* is the most economical approach for analyzing Doppler weather radar signals. If all backscattered energy has no relative motion, the waves coming back to the radar do not change, and backscattered frequency and wavelength is same. Additionally, if we compare the entire wavetrain received from one pulse to a wavetrain from another pulse, we would find no differences. Comparing the wave signatures side by side, we would see that they are perfectly synchronized and the waves themselves will not even show any difference in phase.

Pulse pair processing compares the wavetrain from two or more pulses to detect if a phase shift has occurred, and if so, quantify the shift in order to estimate velocity. Such a shift will be caused by *radial movement* of backscattered energy. The phase shift is relatively easy to measure. However the radar can *only* measure movement along a radial.

4.1.2. ASSUMPTIONS. Doppler radar is often thought of as measuring the speed of the wind within a storm. This is a misconception. Air is a gas and reflects a negligible amount of electromagnetic energy to the radar, so most of the detectable velocity information actually comes from water droplets, ice, rain, and hail. It is assumed that all of these particles are moving with the wind.

Title image
Aliased base velocity image of Hurricane Sandy at Brookhaven, New York on 29 October 2012 at 2221 UTC. Here we are looking at the environmental wind field rather than at the storm-scale level. Winds across the entire image are moving from east-southeast to west-northwest. At this particular configuration and tilt, the product had a maximum unambiguous velocity of 51.5 kt. However a substantial part of the image at the top left and bottom right exceeded this speed by a large margin. Therefore it has "folded over", changing signs and complicating what is essentially a simple two-color velocity field. Proper selection of a de-aliasing algorithm would renders this image correctly with green inbounds on the right side and red outbounds on the left side.

Figure 4-1. Pulse pair processing is achieved by comparing wavetrains from two successive pulses in order to detect phase shifts. The top illustration shows backscatter wavetrain pairs perfectly synchronized, indicating a phase shift of 0° and thus a velocity of zero. The middle shows a phase shift of 90° and indicates a nonzero velocity has been detected in the pulse volume. The bottom illustration shows the pairs 180° out of phase, indicating a velocity equalling the Nyquist co-interval. A stronger velocity would cause a further shift past 180°. If the shift is by 10 more degrees to 190°, the radar will instead interpret it as a -170° shift, flip-flopping the sign of the velocity. This would be an aliased velocity.

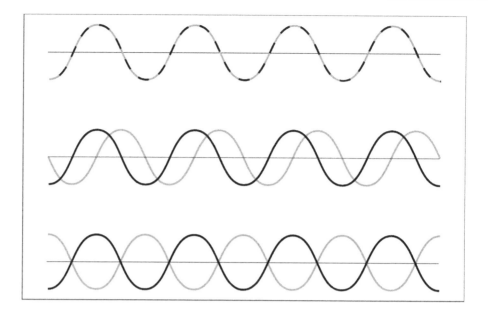

Doppler velocity is also reflectivity-weighted, so within a given pulse volume in a severe storm, the main contributor to the velocity in that volume might be large, wet hail, not small raindrops. The latter are more likely to be moving with the wind.

4.1.3. MAXIMUM UNAMBIGUOUS VELOCITY. An pulse volume contains ambiguous velocity whenever the detected velocity is high enough to cause a phase shift of more than 180 degrees. In other words, if the phase shift is more than half a wavelength, the velocity will be ambiguous. In the case of the WSR-88D, which transmits at a wavelength of 10.7 cm, a phase shift of more than 5.35 cm will cause this to occur.

Now consider that a pulse volume must be sampled twice in order to assess the waveform. Now consider a target with a velocity of 40 m s^{-1}, i.e. 4000 cm s^{-1}. The half-wavelength from a WSR-88D is 5.35 cm. How much time would elapse before the target moved across half of this distance, in other words, a distance of 2.675 cm? Dividing the two on the calculator yields a time of 0.00066875 seconds. The radar must transmit pulses at this rate or faster in order to capture the shift. If we treat 0.00066875 as a PRT, taking the reciprocal gives us a PRF of 1495 Hz. So our PRF needs to be 1495 or higher, otherwise this velocity will be ambiguous. This gives us our equation for maximum unambiguous velocity:

WSR-88D velocities

The maximum unambiguous velocity for the WSR-88D varies depending on which mode the equipment is operating in. The highest value in any of these modes is 64 kt.

$v_{max} = (PRF*\lambda)/4$, where λ is the wavelength

This maximum velocity, v_{max}, is known as the Nyquist velocity, and it gives us the range of unambiguous velocities, 0 to v_{max}, which are known as the

Nyquist interval. Since v_{max} actually applies to motion in either direction, toward or away from the radar, this gives us a range of velocities from -v_{max} to +v_{max} is called the Nyquist co-interval. Any velocities occurring in the pulse volume that fall within the Nyquist co-interval will be considered unambiguous velocities.

The most important aspect of this in operational forecasting is that if we use higher PRFs, velocities within the pulse volume can be higher, because even though the waveform will change more quickly with time over a given distance, the faster PRFs will capture this before the phase shift reaches half a wavelength. So higher PRFs will allow a higher unambiguous velocity (v_{max}).

4.1.4. DOPPLER DILEMMA. In the previous chapter, we showed that decreasing the PRF increased R_{max} and reduced the effects of range folding. However in doing so, a lower PRF reduces v_{max}. A radar operator has the choice of decreasing PRF to increase the maximum effective range of the radar, or increasing PRF to increase the maximum unambiguous velocity. This is the so-called Doppler dilemma.

In order to overcome this engineering problem, the WSR-88D uses a dual-PRF scanning strategy. In the lowest elevations, which are the most critical, the WSR-88D actually uses two specific PRFs to sample each gate twice. The low PRF is known as CS, or contiguous surveillance, and the high PRF is known as CD, or contiguous Doppler.

In the middle elevations, the radar alternates between high and low PRF on each radial. The radar can intelligently compare the two to weed out range-folded data and aliased velocities. On the highest elevations, where second-trip echoes are very unlikely, only a high PRF is used.

4.1.5. DEALIASING. One way to overcome the doppler dilemma is to use a dealiasing algorithm. Since it is a relatively simple operation it can be done using the end user's display software. Dealiasing looks for continuity of velocity along each radial or between adjacent gates. An extreme change may indicate that the velocity has flip-flopped due to aliasing. Dealiasing is not 100% effective and depending on the algorithm in use there can still be aliased areas on the velocity product. The forecaster always has to be on guard for aliased gates and know how to identify and interpret them.

4.1.6. RANGE FOLDING. The problem of range folding has been described in the Fundamentals chapter. While range-folded data can be readily identified and reflectivity products can usually be corrected, this is not true of velocity products. Depending on the radar's operating mode, PRF, and , range-folded data may occasionally obscure large parts of the image. Range-folded data is normally shaded purple, with any velocity data at that location removed. Product legends may use the abbrevation "RF" (range folded).

4.2. Display standards

Since velocity imagery uses a range of values from negative to positive, interpreting the products is not as intuitive as that of reflectivity imagery. There is a very specific meaning to the mathematical sign of a velocity, and to the color scheme that makes up an image. It is essential that forecasters understand this system before attempting to interpret velocity imagery.

All of the information in this chapter onward applies to *single-site* Doppler radar interpretation. The methods for interpreting and displaying dual Doppler radar output are different, and no such networks currently exist in any operational forecast setting.

4.2.1. AXIS. Expressions of Doppler velocity are always radial, that is, parallel to the radar beam. There is no way to detect components of motion perpendicular to the axis. One way to imagine this is to imagine shining a laser pointer inside a perfectly dark room and looking at cigarette smoke floating in the air. The smoke is absolutely invisible except right within the needle-thin laser beam. It is easy to see motion along the laser beam, suggesting movement of the smoke toward or away from you, but it is impossible to determine whether the smoke cloud is moving side-to-side or up-and-down. Naturally you could find more information on its movement by pointing the laser beam to your left, your right, up, and down, and watching the drift on each axis.

4.2.2. SIGN. By convention, positive velocities and "outbound" motions refer to movement away, radially. This motion is either from the radar site (in the case of base velocity) or from the storm motion vector (in the case of storm relative velocity). Likewise, negative velocities and "inbound" motions refer to movement toward the radar site or motion vector. This is a standard that should be committed to memory. A useful mnemonic for this is to remember that the radar site transmits power *away* from the radar; this can be viewed as the normal mode of operation and is associated with a positive sign.

4.2.3. COLOR SCHEMES. Nearly all operational radars worldwide have adopted a convention where red and other "warm" hues are used to depict positive velocity. Likewise, green, blue, and a variety of "cool" hues are used to depict negative velocity. As with the system of positive versus negative velocities, this is also one of the few items in this book that should be memorized and committed to memory.

The amount of saturation and brightness corresponds to the magnitude of the velocity, and in some cases additional blocks of color may be added to further enhance the image. Velocities near zero are depicted using gray colors. Range-folded areas are considered to be degraded and are colored in purple.

Most desktop radar analysis software packages are capable of allowing the user to choose a different color palette and to even create a custom-made palette.

Because proper interpretation of velocity imagery can be critical, this capability is useful for analyzing special situations, such as hurricanes where the color scheme may need to span a wider range of velocities, and for users with color vision deficiencies.

4.2.4. ISOPLETHS. Instead of using colors, it is possible to use isopleths to represent wind velocity. This is often done for the zero-velocity line, which is especially useful for interpreting images. If other lines are drawn, a common convention is to draw inbound velocity isopleths with either a solid or dashed line, and the outbound velocities with the other type of line. While such lines might be described as "isotachs", this term is misleading as such lines do not show an actual wind field but rather the radial component toward or away from the radar or storm motion vector. Only a dual Doppler analysis is able to show true isotachs, and no such network is operationally available. The term *isodop* is occasionally used for Doppler velocity isopleths, though it must be cautioned that this word is still considered by some to be a colloquialism; it does not have an entry in the *AMS Glossary of Meteorology*.

4.2.5. ZERO LINE. One very useful feature in velocity imagery is not the actual values or areas of strong velocities themselves but rather the "zero line". This is the isodop that corresponds to velocities of zero. Using standard velocity color schemes, it can be easily found by following the thin gray band of color.

4.3. Volumetric signatures

Assuming enough scatterers are present in the atmosphere, velocity signatures are produced through much of the available volumetric sample, even in clear and calm weather conditions. Using a single tilt, the velocity signature painted out by the radar in *all directions* provides a signature of the *environmental wind flow*. Some examples of volumetric patterns are shown in Figure 4-2a-f.

When interpreting volumetric signatures, it is especially important to be aware of the conical nature of each radar tilt. * To reiterate this concept, ranges close to the radar site are close to the ground, while ranges near the periphery of the image are at high altitudes. The volumetric velocity signature tends to produce broad regions of colorful negative and positive velocities, but the zero-line is the most useful part of the image because it can be interpreted to be a location where wind flow is perpendicular to the radar site. In fact, wind arrows or wind barbs can be drawn along the zero-line that are perpendicular to the radar site.

Color vision deficiency

Many weather websites that provide radar information give little or no control over the color scheme. This can cause significant problems for those who are "color blind", especially when interpreting critical products like velocity.

Visolve is a Windows & Mac program that can be used to enhance colors to make them more easily viewable. It is free for personal use and available for Windows and Mac OS X. Available at: http://www.ryobi-sol.co.jp/visolve/en/

There is also the ColorBlindExt add-on for the Firefox browser that changes images and text to make it more accessible to those who have color vision deficiencies. Available at: https://addons.mozilla.org/en-US/firefox/addon/colorblindext/

Likewise, for those designing color palettes, the Color Oracle program can be used to show what an image might look like to the color vision impaired. It is available for Windows and Mac at: http://colororacle.cartography.ch/

Figure 4-2a. Volumetric radial velocity image (radar site is at center and furthest available range is on edge of the circle) assuming a homogenous wind field with winds blowing from the south at 60 knots throughout the entire volume. The zero isodop is drawn as a dashed line. It can be seen that north and south of the radar site (directly into and away from the wind), very strong inbound and outbound velocities are occurring at all levels.

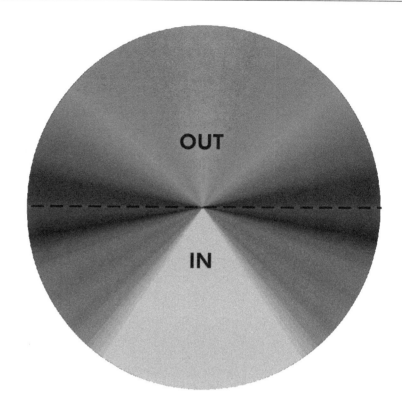

Figure 4-2b. Volumetric radial velocity image assuming a homogenous wind field with winds blowing from the west at 60 knots throughout the entire volume. Again, the zero isodop is drawn as a dashed line. The strongest winds are found east and west of the radar site, with colors corresponding to either -40 or +40 knots along this axis..

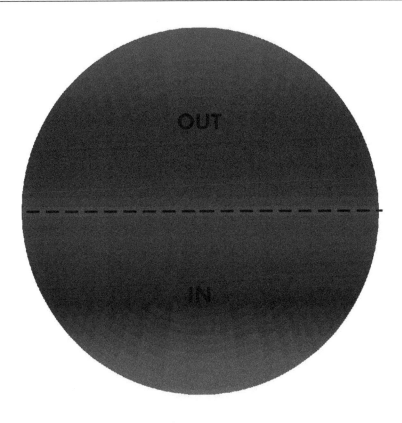

Figure 4-2c. Volumetric radial velocity image in an environment with containing weak speed shear. Winds are calm at the surface, increasing to southerly at 20 kt at the highest elevations. Note that the zero isodop is straight, indicating north-south flow at all levels.

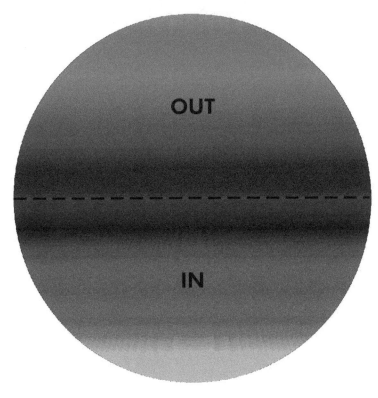

Figure 4-2d. Volumetric radial velocity image in an environment with containing strong speed shear. Winds are calm at the surface, increasing to southerly at 60 kt at the highest elevations. The zero isodop is straight, even in the presence of stronger winds.

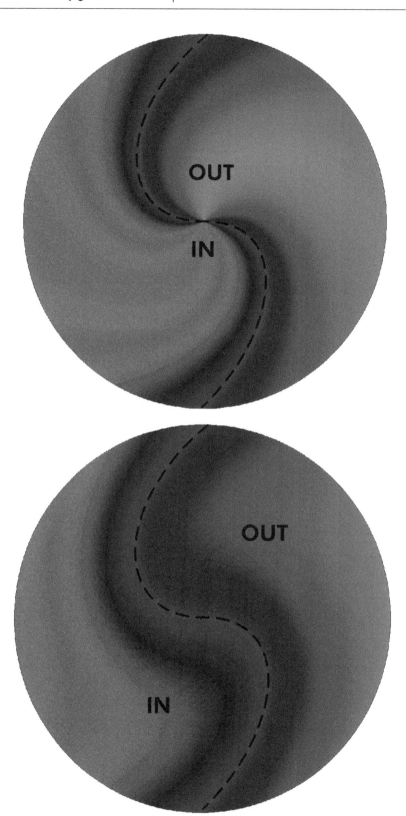

Figure 4-2e. Volumetric radial velocity image in an environment with no speed shear but with directional shear. This creates a prominent "swirl" effect, with curvature of the zero isodop. Note that the maximum outbound and inbound velocities are not concentrated near the surface but are simply the same magnitude throughout all levels. In this example, winds are from the south at 40 kt at the surface, changing to west at 40 kt at the highest elevations.

Figure 4-2f. Volumetric radial velocity image in an environment with both speed shear and directional shear. The winds in this example are southerly at 1 kt at the surface, increasing to southwesterly at 20 kt in the mid-levels and westerly at 40 kt aloft. There is less of a "pinched" look to the velocity field near the radar site, owing to the weak wind field near the surface, but the same strong velocities in the previous image are found along the edge of the volume in the higher elevations.

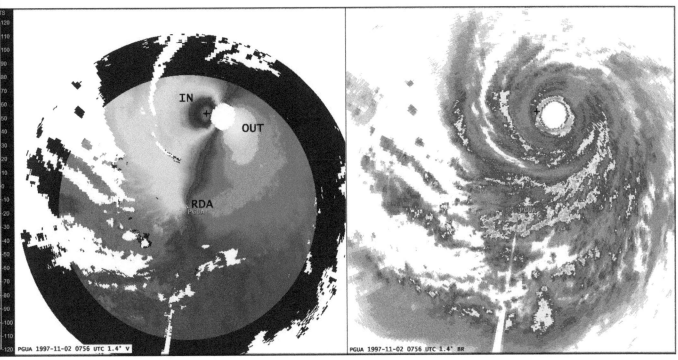

Figure 4-3. Radial velocity image of Super Typhoon Keith off the north coast of Guam on 2 November 1997. The 0.5° tilt was affected by beam blockage, so 1.4° is used here. The radial velocity product near the typhoon's eye indicated inbound winds (marked by the cross) of 126 kt (145 mph). This reading is 68 nm from the radar, so the eye is actually being sampled at an elevation of 13,000 ft MSL. The 0.5° tilt, which was at an elevation of 6,000 ft MSL, indicated similar wind profiles. The outbound winds on the other side showed values of up to 145 kt (167 mph) at both tilts! Even at the 4.3° tilt, which intersected the eye at 33,000 ft, radial winds were over 100 kt on both sides! However at this tilt the cyclonic rotation dipole seen here rotated slightly counterclockwise around the eye, indicating a transition to the expected divergent outflow.

4.4. Localized signatures

Localized signatures are not radar-wide, most often on the same scale as a storm or even smaller. These yield information not on the environment but on the small-scale weather phenomena occuring with it.

4.4.1. COUPLETS. The concept of a "couplet" is extremely important to understanding small-scale circulations. The couplet is a name for any small-scale dipole, in which one pole has strong inbound velocity and the other pole has strong outbound velocity. In some cases the couplet can be masked by improper use of base velocity, or in the case of storm-relative velocity products, by the RPG or display software using an improper storm motion vector. If storm-relative velocity is being used and the forecaster is able to adjust the storm motion vector, it is best to find a setting that "balances" the velocities of the couplet but still properly represents the overall motion of the storm or the mesocyclone.

The size and intensity of a couplet yields information about the process that is taking place. Software that allows the use of a cursor to measure the size of the dipole will yield the diameter of the circulation.

4.4.2. SHEAR AND ROTATIONAL VELOCITY. There are two expressions of velocity: shear and rotational velocity. Shear is the difference in velocity between the maximum and minimum velocities in the couplet, so if there is +60 kt outbound and -40 kt inbound, the shear is 100 kt. Rotational velocity, on

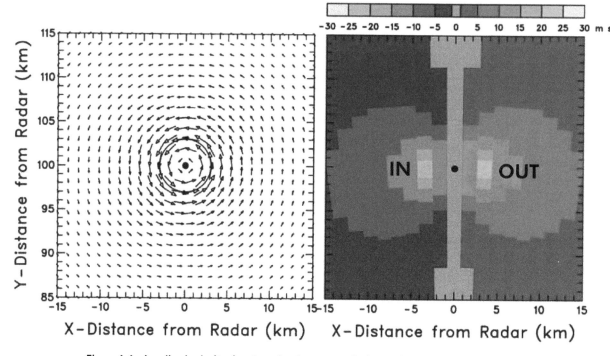

Figure 4-4a. Localized velocity signature showing pure cyclonic rotation. Instead of depicting all directions from the radar, as in the environmental wind sets, here we are looking at only a localized area. The radar is to the bottom of the page in all images in this set (i.e. to the south). This example shows that the axis of the "couplet" is perpendicular to the radar, with the configuration shown here. *(This graphics set from A Guide For Interpreting Doppler Velocity Patterns, Rodger A. Brown and Vincent T. Wood, National Severe Storms Laboratory, 2007)*

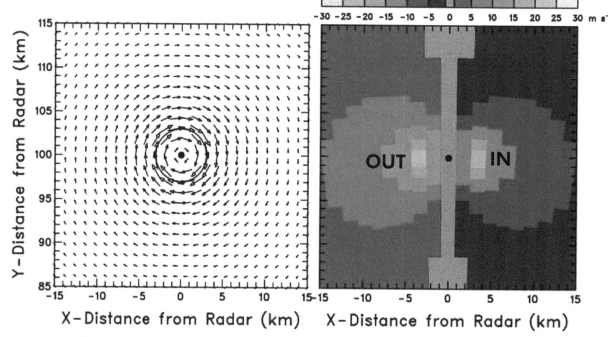

Figure 4-4b. Localized velocity signature showing pure anticyclonic rotation. Same as above but the sign is reversed.

4 | VELOCITY   73

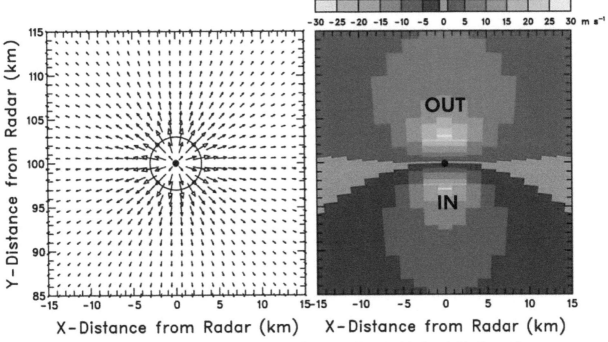

Figure 4-4c. Localized velocity signature showing pure divergence. The axis of the "couplet" in divergent/convergent signatures is parallel to the radar. With divergence, the inbound portion is the area that is closest to the radar.

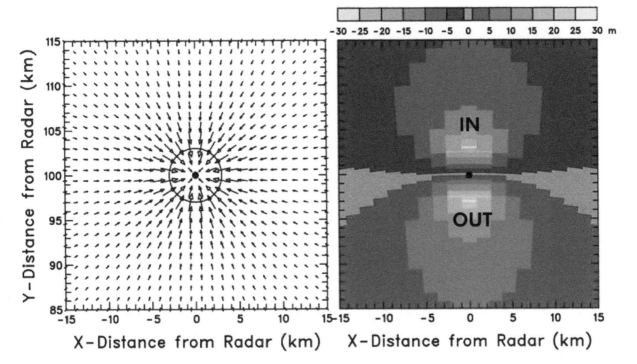

Figure 4-4d. Localized velocity signature showing pure convergence. Again, the axis of the "couplet" in divergent/convergent signatures is parallel to the radar. With convergence, the inbound portion is the area that is furthest away.

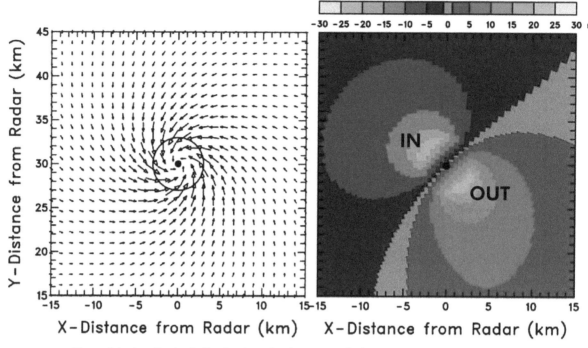

Figure 4-4e. **Localized velocity signature showing pure cyclonic convergence**, as might be detected in a mesocyclone or tornadic vortex signature. This is a combination of the cyclonic and convergence signatures, with the couplet axis diagonal to the radar..

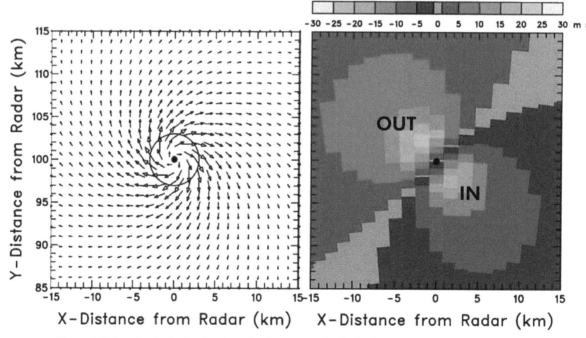

Figure 4-4f. **Localized velocity signature showing pure anticyclonic divergence**, as might be detected in a downburst. This is a combination of the anticyclonic and divergence signatures, with the couplet axis diagonal to the radar..

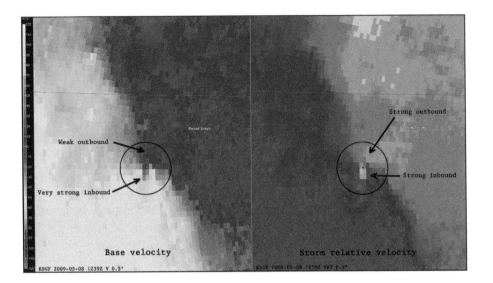

Figure 4-5. **Using storm-relative velocity** to bring out the signature of a brief tornado on the leading edge of a derecho in southwest Missouri. The storm relative velocity product more effectively shows the tornadic rotation, indicating 45 kt of storm-relative outbound velocity on the north side and 45 kt of inbound on the other side.

the other hand, is basically tangential velocity and is simply the average of the absolute inbound and outbound velocity in the couplet. For the above example the rotational velocity would be 50 kt. Tornadoes are usually described in terms of shear, while mesocyclones are described according to rotational velocity.

4.4.3. RADIAL VELOCITY. The WSR-88D base velocity product is radial velocity. This is simply the velocity field at any given tilt exactly as detected, without any correction for storm motion. In other words, it is a ground-relative velocity. This product tends to be the default velocity product seen on weather websites, mobile devices, and display software.

4.4.4. STORM RELATIVE VELOCITY. Radial velocity uses the stationary Earth's surface as a frame of reference. Small-scale circulations and entire storms, however, almost always move with respect to the ground and this can distort signatures on base velocity products or make them difficult to analyze. For example, a storm moving away from the radar at 50 kt and containing rotation with 20 kt of shear will show velocities of +40 and +60 kt embedded in a storm that shows about +50 kt of motion. It is much more useful to subtract out the component of storm motion to provide a storm-relative framework.

The Level III storm-relative velocity product is computed using a vector produced at the RPG. In many cases, this is sufficient to analyze a storm. However in other cases, particularly in situations like storm split events where cells are moving in different directions, a forecaster may want to select a specific storm motion vector. This can be performed on Level II and Level III base velocity products using certain desktop radar viewers equipped with their own storm motion vector input. An example of such a program is GRLevel2 and GR2Analyst.

Strong mesocyclones
What is a strong mesocyclone? This is difficult to answer since it depends on circulation type and diameter, range from the radar (beam broadening), and other factors.

Mesocyclone (3.5 nm diameter)
STRONG if rotational velocity is:
>45 kt near the radar
40 kt at 60 nm
35 kt at 100 nm

Mesocyclone (1.0 nm diameter)
STRONG if rotational velocity is:
>40 kt near the radar
30 kt at 60 nm
20 kt at 100 nm

Storm-relative velocity should be used for most severe weather forecasting purposes, but should not be used when a forecaster is using the lowest tilt to assess downburst wind potential. In this case, base velocity will more accurately represent the winds experienced at the surface.

4.4.5. MESOCYCLONE. A mesocyclone is normally manifested by a strong rotational couplet with a diameter of 2 to 8 miles, while tornadoes are often involved in strong rotational couplets with diameters of less than 2 miles.

4.4.6. TORNADIC ROTATION. The term "gate to gate shear" was frequently used in the 1990s and early 2000s to quantify the strength of the tornado by assessing the intensity of rotation between two adjacent pulse volumes. Owing to the much higher resolution of upgraded WSR-88D equipment and algorithms, high intensities are now more likely to be spread out across many gates rather than concentrated between two gates. Because of this, forecasters should not be looking specifically for intense gate-to-gate shear. When it is seen and shows consistency, though, a tornadic circulation or some other localized circulation may be indicated.

A more meaningful quantification of shear is the apparent diameter of rotational couplets together with the magnitude of shear, e.g. "one mile wide circulation with 75 kt of shear". Generally intense circulations with a diameter of 1 mile or less are tornadic, while larger couplets are associated with the mesocyclone. There is no specific value of diameter and magnitude that differentiates between the two, because of the variety of circulations that may occur and because gates have a variety of volumes and heights, depending on the range and beam width. In some cases, the beam may overshoot the tornadic circulation altogether.

4.4.7. DOWNBURSTS. A downburst is indicated on radar by the presence of a strong, localized divergence signature on storm-relative velocity. The forecaster must also switch to radial velocity in order to measure ground-relative winds. However, only the radial component of this wind can be obtained, so actual surface winds may be higher. Additional evidence that supports a downburst include increasing reflectivity at the downburst center, caused by the precipitation cascade, and high intensities or high values of vertically integrated liquid (VIL) overlying the downburst center.

4.4.8. BOW ECHOES AND DERECHOS. A bow echo is a downburst with some degree of organization, often with clearly-identifiable mesoscale structures. The rear inflow notch was mentioned in the Reflectivity chapter as a signature that might mark the presence of a developing downburst. Base velocity and storm-relative velocity products can help to evaluate these areas further.

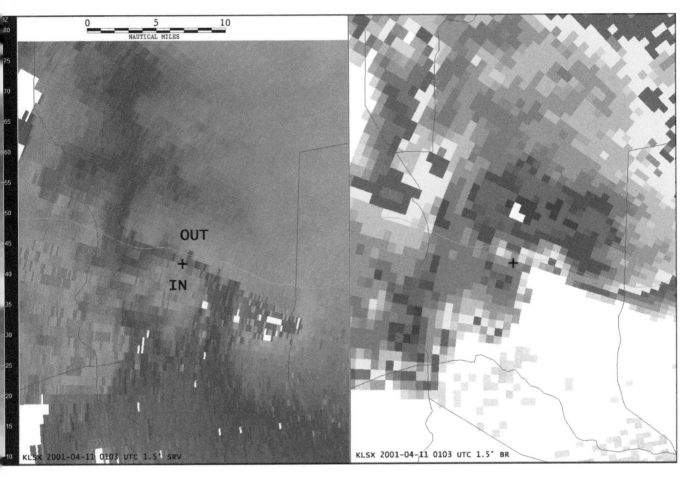

Figure 4-6. Typical appearance of a tornadic HP storm as seen on storm-relative velocity (left) and reflectivity (right). The radar is to the right and slightly below. At the crosshair, marked with a bold + symbol, storm relative velocity detected cyclonic shear of 73 knots between adjacent bins. In this case, the diameter of the circulation is about 0.5 nm. This is embedded in larger scale rotation, the mesocyclone, with a diameter of 4 nm and about the same magnitude of shear. This was a storm on the evening of 10 April 2001 near Jonesburg MO. Imaged with GR2AE and annotated by the author.

The base (ground-relative) velocity product should be used to determine surface-based wind speeds and evaluate damage potential. The caveat is that the beam will likely overshoot the surface winds to some extent. Moreover, Doppler velocities are radial, so wind speeds will only be representative if the estimated wind direction is believed to be directly toward or away from the radar. If the wind direction has a perpendicular component, it may be necessary to adjust the wind speeds upward or select a different radar site, if one is available.

4.5. Spectrum width

Spectrum width measures the range of velocities within a pulse volume. It is a measure of *diversity* of velocity, and is expressed as a standard deviation. The higher the spectrum width number, the greater the diversity. High numbers indicate a mix of hydrometeors with different mass and sizes, some moving more more uniformly with the wind and others less so, for example rain mixed with

Figure 4-7. An extreme closeup (note the distance scale) of storm-relative velocity near a radar during a tornado event. Isopleths corresponding to velocity ("isodops") have been drawn manually to emphasize the amount of outbound (solid) and inbound (dashed) velocity. GR2AE and Level II data was used to carefully set a storm motion that "balanced" the couplet. This shows 120 kt of shear across a half-mile diameter, a strong indicator of a tornado. This was the 10 May 2010 tornado near E. Lindsey & 72nd Street in Norman OK shortly before producing EF4 damage.

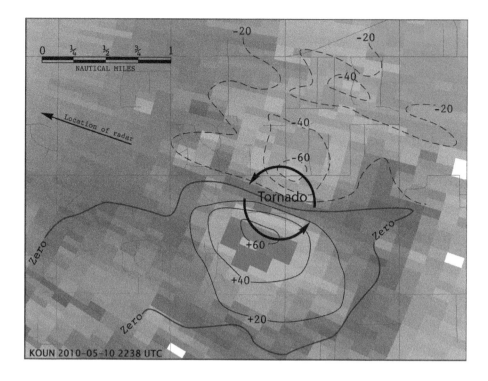

Figure 4-8. A tornadic storm at long distance has a much different presentation, especially without the benefit of "super resolution". This is the Fort Smith tornado on the evening of 21 April 1996, viewed with the Tulsa (KINX) radar, the closest available site. The 0.5° base reflectivity is shown at right, revealing signatures of a classic supercell. The radar is 72 nm from the radar, intersecting it at 7000 ft; compare the distance scale with that in the example above. Estimated boundaries are overlaid with dashed lines. The 0.5° storm-relative velocity at left is configured with the 55°/35 kt mean vector of the mesocyclone. It shows 115 kt of "gate-to-gate" shear. Forecasters should not be looking specifically for strong gate-to-gate shear, since today's radars have a fine enough resolution where shear can be diffused across multiple gates, but when it occurs, especially at far distances like this, it is a strong indicator of tornadic winds. Deficiencies in radar coverage in southeast Oklahoma and west Arkansas combined with the lack of a tornado warning on this storm prompted the construction of a WSR-88D radar at Fort Smith (KSRX) in 1998.

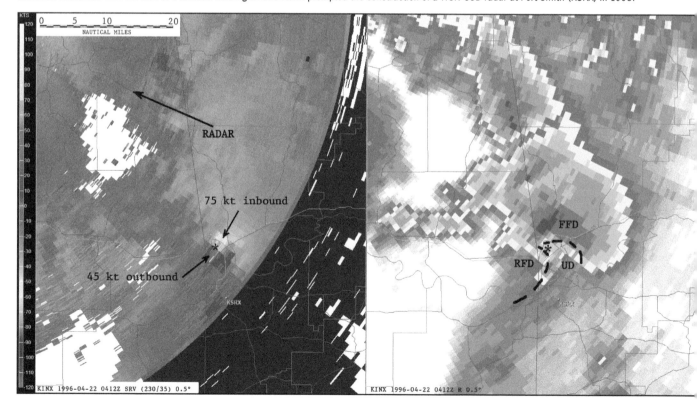

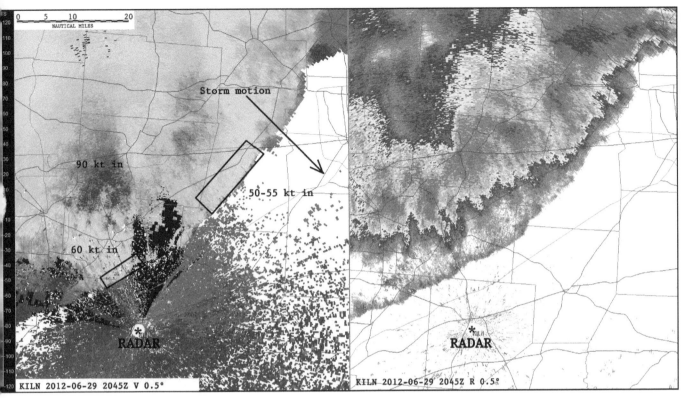

Figure 4-9. The derecho of 29 June 2012 as seen by the Wilmington, Ohio radar. The radial velocity product is shown on the left. As the storm is moving to the southeast, the band of 60 kt winds within the outflow close to the radar represents a good estimate of wind gusts expected at the radar site. Further northeast along the line, the outflow amazingly shows 50 kt of inbound wind, even though the system is moving southeast. It can be safely assumed the winds in this region are probably stronger, probably 80 to 100 kt. However it must also be remembered that there, the storm is being sampled at 2500 ft AGL, instead of the 500 ft level close to the radar site. Tracking the outflow boundary itself can also give a minimum estimate of wind speed.

hail. Low numbers indicate a distribution of particles in a pulse volume with similar size, mass, and composition.

Spectrum width is especially noteworthy for being an underdog at the operational level. Only a handful of journal articles have explored its operational use, and many Internet radar sites do not offer it at all. However in recent years, especially with the increased resolution of WSR-88D products, it has proven to be exceptionally useful for locating outflow boundaries, particularly those embedded in precipitation. It also shows reveals the high diversity signatures in the tornado debris cloud, owing to the wide range of materials with different aerodynamic properties being lofted into the tornado circulation. Spectrum width can not only discriminate outflow boundaries embedded in precipitation, but it can also help confirm the presence of embedded and rain-shrouded tornadoes.

4.6. Wind profile products

The WSR-88D has algorithms for determining the volumetric wind at different levels. These are an average of winds throughout all radials, so corruption is likely to occur when winds within the radar volume are not uniform.

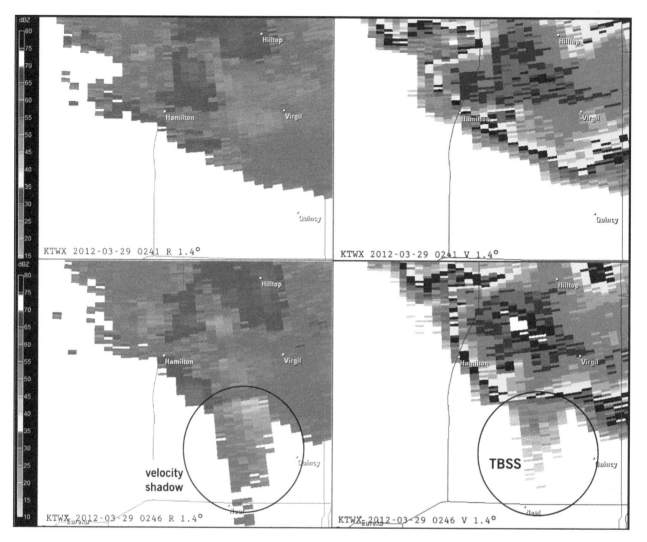

Figure 4-10. Appearance of three-body scatter spike (TBSS) and velocity shadow on velocity (left) and reflectivity (right) imagery. The features in the bottom imagery appeared five minutes after the top set of images, and disappeared just as quickly as they appeared. An unaware forecaster may classify this as a dangerous circulation, and it is furthermore possible that WSR-88D or third-party algorithms may identify a tornado or mesocyclone in this area.

4.6.1. VELOCITY AZIMUTH DISPLAY (VAD). The Velocity Azimuth Display is not seen operationally, but should be reviewed by forecasters since it forms the basis for the widely-used VAD Wind Profile. This is essentially a plot of azimuth versus velocity at a given tilt.

4.6.2. VAD WIND PROFILE (VWP). The velocities found throughout the entire volume can be analyzed to provide an estimate of the winds within the atmosphere above the radar site. It is said this is an estimate because the radar antenna is not aimed vertically, but rather, many different directions and distances are analyzed, some of these distances 100 miles away or more, and the information from each of those bins is assumed to represent the winds above the radar site. The VWP product is provided not as a map but as a chart, with the horizontal axis representing time and the vertical axis representing height above ground level. So the VWP not only provides wind information aloft but provides a time series. Wind samples which do not match the representative "curve" fitted to the wind field are considered to have a high RMS error, or low confidence, and are usually plotted in shades of yellow or red.

Review Questions

1. By reducing the PRF we increase the maximum unambiguous range of the radar. What does reducing the PRF do to the maximum unambiguous velocity?

2. What is the wavelength of radiation from a WSR-88D radar? Express this as centimeters, meters, and kilometers.

3. When a WSR-88D is operating at a PRF of 1304 Hz, how fast can a target move so that it is sampled twice within a half wavelength and thus its velocity is not ambiguous?

4. What is the name for the product in which storm motion is subtracted from radial velocity?

5. Describe the mathematical sign for inbound and outbound velocities, and their standard display color.

6. What is the difference between shear velocity and rotational velocity?

7. What might be signified by intense, localized divergence within a downdraft area?

8. When the maximum unambiguous velocity is exceeded at a bin, what happens to its values?

9. Which product shows the diversity of velocities at each bin?

10. There is a ring of missing velocity information around a radar site. What is this likely to be?

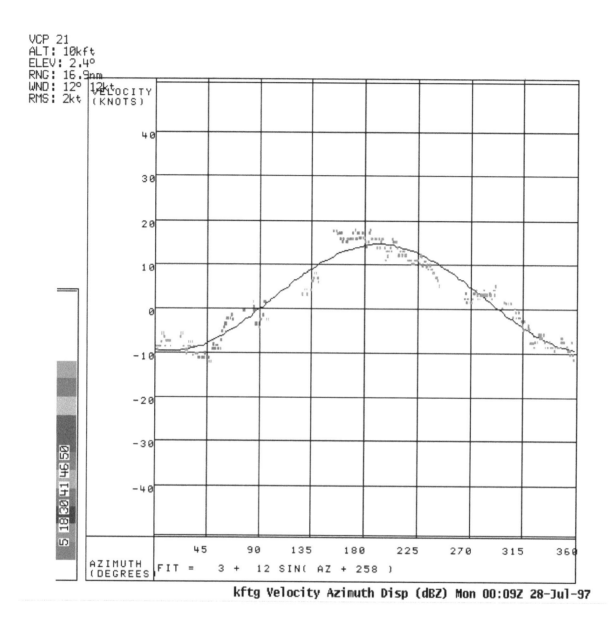

Figure 4-11. Velocity azimuth display showing, for a given tilt, azimuth versus the velocity detected at a specific height (range). Since all of the detected wind information is the component of wind relative to the radar site, it is possible to fit a sine wave to all the data. The horizontal position, or phase, of this wave indicates the wind direction, and the amplitude corresponds to the wind speed. The further a wind sample lies from the curve, the greater the RMS error and the more uncertainty there is with that sample. It can be seen that anomalous winds in a certain sector, such as a front to the northwest containing different wind directions, would contribute samples with high RMS errors and would make it more difficult for the processor to fit the curve.

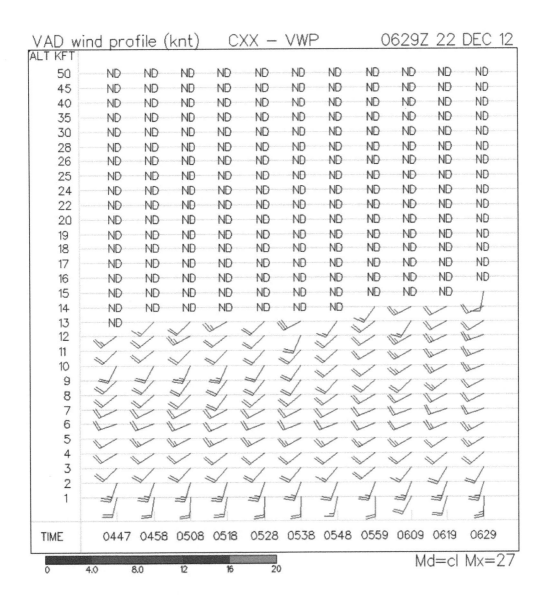

Figure 4-12. VAD wind profile (VWP), which is simply a plot of wind direction and speed with respect to time (horizontal axis) and height (vertical axis). The barbs are standard meteorological wind barbs, pointing into the wind and with each long feather indicating 10 knots. The marks ND mean "no data".

Figure 4-13. Perspective is everything! In this example the Norman, Oklahoma downburst event of 14 June 2011 is shown, along with a special example of velocity products. We see the storm from two WSR-88D radars: the KTLX (Twin Lakes) network radar, and the KOUN (Norman) research radar. They are 11 nm (20 km) apart.

The 0.5° base reflectivity product (below) from KTLX is shown for reference purposes. One very important severe weather indicator is the strong intensities at the highest tilts (dashed line, representing 60 dBZ at 19.5°). This is strongly displaced ("tilted") from the location of the storm in the low levels and is a prominent severe weather marker.

The 0.5° base velocity product from the network KTLX radar (top right) seems to show an east-west axis for the downburst and the areas of highest wind. Since these velocities are outbound/inbound components of motion, a wind flow pattern shown by the arrows is suggested. This might lead a forecaster to issue warnings for the area south of the downburst axis.

However a look at the same base velocity from KOUN, 10 miles to the southwest (bottom right), at almost the very same time, shows a more northeast-southwest or even north-south axis. Looking at this, a forecaster might be inclined to warn of high winds east of this axis rather than to the south.

The illusion here is that the axis of high winds in both cases is broadside to the radar. The radar only measures radial winds, that is, the component of motion to and from the radar. If the downburst is not blowing directly along the radial, then it will show up as a weak velocity and may not stand out at all, even while damage is being produced. The forecaster must use all available products to judge the extent of the outflow boundary, use time steps to estimate the direction of the wind, and anticipate areas where the radial component is likely to be weak, for example if the radar is perpendicular to the track of a bow echo.

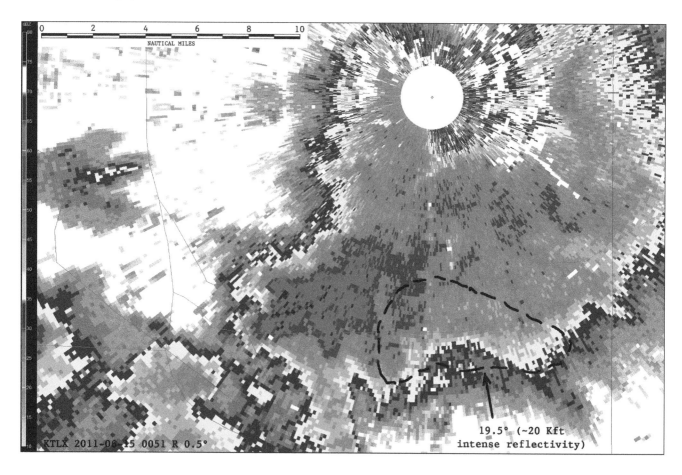

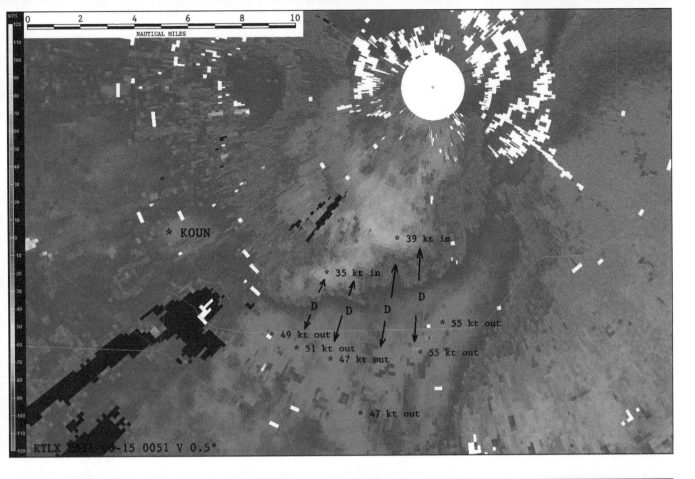
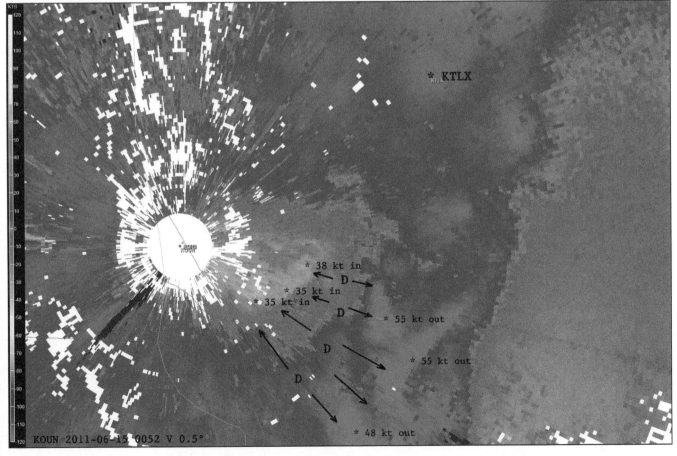

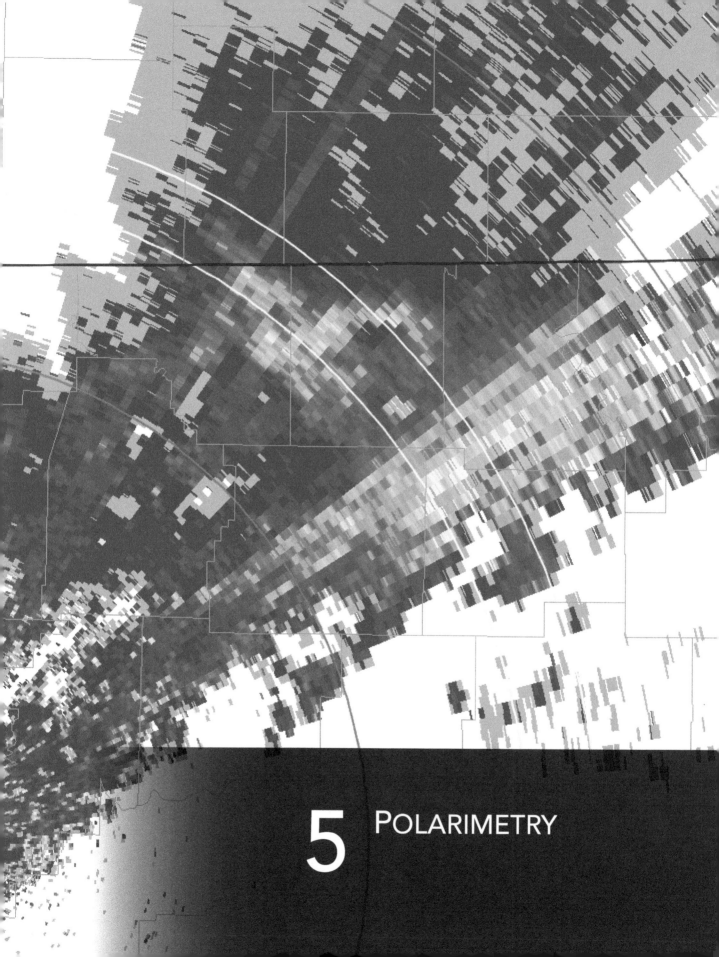

5 POLARIMETRY

5 | Polarimetry

Most readers are familiar with the image of a television antenna mounted on a pole, an iconic symbol of post-WWII suburbia. Why are these television antennas always mounted level with the ground, and not sideways on the pole? An hour of frustrating work outside will certainly reveal the answer, but it's easier to explain it in terms of electromagnetic energy. Television broadcasters cause their signal to radiate in a way that the electric field oscillations of the electromagnetic waves are flat with respect to the ground, in other words, horizontally polarized. Mounting the rooftop antenna to be flat with respect to the ground is a way in which, without thinking about it, we have matched the antenna's polarization to that of the broadcast source.

For many decades, operational weather radars simply transmitted and received radar energy in the same phase, usually horizontally. However, since the 1970s researchers have been experimenting with *polarimetric radars*, a technology which compares and contrasts electromagnetic energy received and transmitted with different polarities. Using this technique it is possible to get a substantial amount of additional information from inside clouds, storms, and weather systems. By 2003, this technology had matured sufficiently to where the United States National Weather Service approved the conversion of its entire WSR-88D radar network to dual-polarimetric ("dual pol"). Most of this work was done in 2011 and 2012.

Though polarimetric radars are cutting-edge technology, they are still an add-on to conventional radars and all of the material in the preceding chapters still applies. This chapter strictly deals with the polarimetric measurements of weather radars in an operational setting. Since some of the polarimetric terminology is rather wordy, many readers may find the symbolic use of variables to be more intuitive and immediately recognized, so these will be used throughout the text to improve the readability. These variables are also frequently encountered in scientific papers and presentations without much context. So it is recommended to memorize the variables as they are presented here.

5.1. The basics of polarimetry

The objective of polarimetric radar is to measure the electric field of backscattered radiation from a single pulse volume in both the vertical and the horizontal plane. While this sounds simple in principle, it is not cost-effective to build a radar that transmits and receives in both of these planes, since this requires an expensive electromagnetic switch and increases the time needed to sample the pulse volume.

One compromise is called slant-45, in which a pulse is transmitted which is polarized 45 degrees from the vertical. This is the method used by polarimetric WSR-88D radars. The radar receives the backscattered radiation and separates it passively into the vertical and horizontal planes, making its reception measurements in this manner. The advantage of this technique is that only

Polarimetric calibration
Overall, polarimetric radars must undergo a much more rigorous calibration process as compared to conventional radars. The components are also highly sensitive to temperature.

Title image
Correlation coefficient product along a weak squall line early in the morning hours of 4 December 2012 in northwest Arkansas. Several rings are also overlaid, indicating the melting layers. Most of the brighter areas indicate lowered correlation coefficients due to the presence of wet snow and other mixed phases where the beam intersects melting levels.

Polarimetry

Polarimetric radars can be divided into two basic characteristics of electromagnetic polarization: linear and circular polarization. Circular polarization rotates the polarization with respect to time. These types of radars are more expensive but they can be very effective at differentiating non-meteorological targets such as airplanes from heavy rain. Linear polarization compares the polarization along two or more fixed axes. Most of the existing advanced operational weather radars, including the WSR-88D, strictly sample the horizontal and vertical axes, thus they are known as "dual-polarization" or "dual-pol".

Difference reflectivity

Though differential reflectivity is by definition expressed as a ratio, $10\log_{10}(Z_H/Z_V)$, it is also possible to express it as a difference, $10\log_{10}(Z_H-Z_V)$. This yields a parameter called difference reflectivity (note the slight spelling change). This is not provided as part of the WSR-88D dataset, but it is occasionally used in research radars for studying rain/ice mixtures.

a single pulse is needed to sample a pulse volume. The problem is that since backscatter power is split between the two components, there is signal loss.

An important note is subscripts. These are used in all of the parameters and indicate what type of polarity is used. For example, Z, which is reflectivity, is written as Z_{HH} or simply Z_H if we are referring to "normal" conventional reflectivity. This indicates the radar is both transmitting and receiving in the horizontal plane. This yields no polarimetric characteristics. With a polarimetric radar, we can construct images showing reflectivity in the horizontal plane (Z_H) and the vertical plane (Z_V). However instead of doing this, it is more useful to obtain the ratio between these two phases (Z_{HV}). This is an expression of *cross-polar power* and is a polarimetric measurement.

5.2. Differential reflectivity (Z_{DR})

Differential reflectivity (Z_{DR}) is perhaps one of the easiest polarimetric parameters to understand. The radar measures backscattered radiation in the horizontal plane (Z_H) and the vertical plane (Z_V). At any given gate, the two values can compared in the form of a ratio, Z_H/Z_V. This ratio Z_H/Z_V can also be treated as a mathematical fraction. If equal amounts of power are returned, the result is simply 1. But if more power is received in the horizontal plane than the vertical, the term Z_H becomes larger, yielding a value greater than 1. On the other hand, if the opposite is true, Z_V is larger and we get a value less than 1.

One twist is that differential reflectivity is not defined simply as Z_H/Z_V, but $10\log_{10}(Z_H/Z_V)$. It is not important to know the math, but simply that when the ratio yields 1, the result will actually be 0. So zero becomes the dividing line, with positive values indicating that Z_H is higher and negative values indicating that Z_V is higher.

How does all this apply to weather? First let's imagine that a gate (sample pixel) consists entirely of perfectly spherical raindrops. The radar might detect 30 dBZ using the horizontal phase and 30 dBZ for the vertical phase. This is because the raindrops are spherical, so no matter which way we "turn" the antenna, the raindrop geometry will look the same and the same amount of power will be reflected. So this gives us a ratio of 1-to-1, and 1/1 equals 1. Since $10\log_{10}(1) = 0$, the differential reflectivity product will show 0 dBZ.

Now let's take an extreme example of large raindrops, the same ones which splatter loudly when they impact the ground, and consider what happens if they precipitate out of a cloud. These large drops fall at a high velocity, and as they fall they encounter drag as air molecules beneath resist the large surface area of the drops. This causes them to flatten out as they fall, taking on an *oblate* shape, with the same shape and orientation as we might imagine with a slowly descending flying saucer. If we interrogate them with a polarimetric radar, we might find 40 dBZ of reflectivity using horizontal polarization, but a much lower 10 dBZ using vertical polarization. This gives a ratio of 4/1, which equals 4. Using $10\log_{10}(4) = 6$, so the differential reflectivity product will show 6 dBZ.

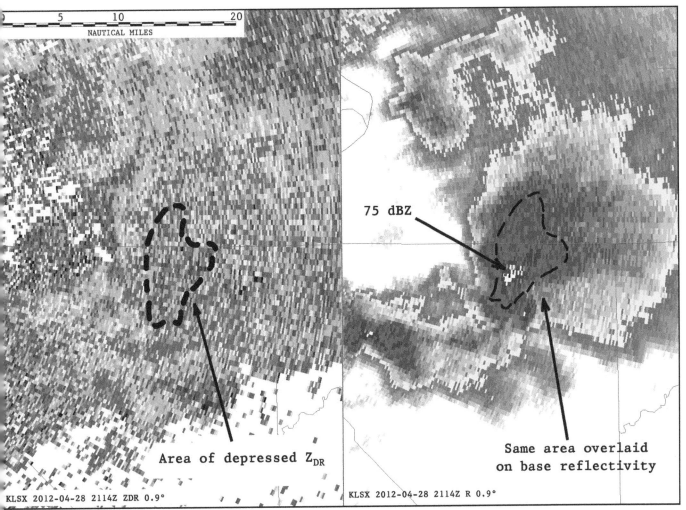

Figure 5-1. Damaging hailstorm east of St. Louis on 28 April 2012. The right frame shows conventional base reflectivity, while the left frame shows differential reflectivity. The core of the storm shows purple and blue areas corresponding to between -1 and 0.5 dBZ, much lower than the values of 4 to 7 dBZ that would be expected with heavy rain. At the time, 4:14 pm CDT, the core was just north of O'Fallon, IL, and was producing hailstones 2 to 2.5 inches in diameter (between golf ball and baseball size)..

On the other hand, solid particles like hail behave differently. While many hailstones are spherical, those that are oblate will tend to "weathercock" into the wind as they fall, yielding a *prolate* shape like a football falling with its end pointing at the ground. This gives a much larger vertical profile compared to the horizontal one. Such a pulse volume might give a power of 60 dBZ using vertical polarization but 30 dBZ using horizontal polarization. This gives a ratio of 3/6, which equals 0.5. Since $10\log_{10}(0.5) = -3$, the differential reflectivity product will show -3 dBZ.

Other targets that have unique differential reflectivity signatures. Insects have a high Z_{DR} since their bodies tend to have a horizontal orientation while in flight.

So differential reflectivity provides a useful measure of the *shape* and *orientation* of particles, and thus whether they are liquid or solid, and if they are liquid, what the approximate drop size is. The higher the differential reflectivity, the more likely the backscatterers are liquid and the greater the drop size. The

CC: What is high? What is low?
Shown here is an approximate listing of correlation coefficient ranges:
0-0.6   Very low
0.6-0.94 Low
0.95-0.96 Medium
0.97-1   High

Ice particle types

Planar
Single dendrites
Plates
Stellars
Aggregates of the above

Columnar
Columns
Bullets
Needles

Quasi-spherical
Ice pellets
Graupel
Hail

lower the differential reflectivity, the more likely the backscatterers are solid, and the more likely they are to be hail.

5.3. Correlation coefficient (CC, rho, ρ_{HV})

The cross-correlation coefficient (ρ_{HV}, or simply rho) compares the consistency between backscattered power in the vertical and horizontal planes. If the orientation of scatterers does not change from pulse to pulse, the CC value will be high.

The CC product is most useful for differentiating hydrometeors, which have high ρ_{HV} values, from non-hydrometeors, which have low values. For example, a volume containing birds and insects will have a very wide variety of shapes, and the result will be a correlation coefficient of less than 0.90. The value of 0.90 tends to differentiate almost all non-meteorological targets from those that are meteorological, with the exception of large, wet hail which disperses energy through Mie scattering and large, wet snowflakes. Most other meteorological backscattered energy has a correlation coefficient of above 0.90.

More importantly, however, it is used to identify regions which have a large diversity of precipitation types, shapes, orientations, and sizes. The exact ρ_{HV} value is dictated by the diversity of *shapes, oscillation, wobbling, and canting* of the precipitation particles. Low values are likely to be caused by non-hydrometeors and particles that change their orientation significantly between pulses. Medium values (0.6 to 0.95) are associated with a high diversity ice and liquid precipitation, irregular shapes, many different orientations, and large hail. High values are associated with a uniform precipitation field and particles that do not change their orientation significantly between pulses.

Rain, ice pellets, graupel, and snow all produce values of greater than 0.95. However in a pulse volume in which liquid precipitation is mixed with solid precipitation, the correlation coefficient will drop below 0.95. Drizzle produces values less than 0.90 because of poor signal-to-noise ratio. Hail tends to have a relatively low correlation coefficient due to the frequent occurrence in a mixed solid-liquid volume and due to dispersion of energy through Mie rather than Rayleigh scattering.

In tornadic situations, correlation coefficient can be used to identify tornado debris, which will have low values. This type of use may be difficult since most tornadoes occur on the edge of the storm, where there may be a lack of scatterers or very noisy correlation coefficient fields, and hail, which has low CC values, may be wrapped into the tornado. So this technique is most likely to be useful primarily in embedded tornadoes with no significant hail. It follows that this would be likely in low-instability, high-shear weather patterns.

The engineering quality of the radar has an effect on the cross-correlation coefficient. A cheaper radar with a marginal antenna, transmitter, and signal processor will tend to produce lower values of ρ_{HV} overall.

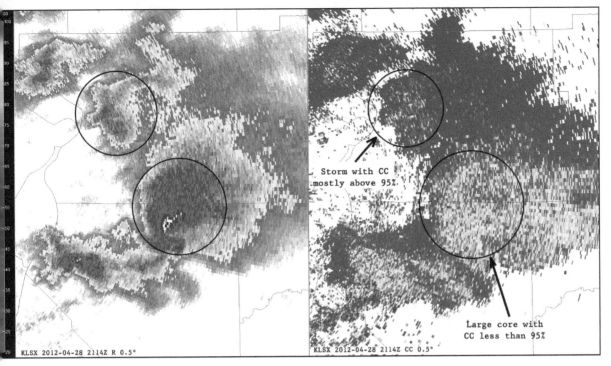

Figure 5-2. Correlation coefficient of a nonsevere storm and a hailstorm, same event as in Figure 5-1. The conventional base reflectivity is shown in the lleft frame. Correlation coefficient is shown in the right frame. The default palette in GRLevel2 AE was replaced with the AWIPS palette, which gives less noisy results. Red colors such as those in the circled storm to the northwest correspond to values of 95% or more, while other colors indicate values of less than 95%. The only types of precipitation which will produce returns less than 95% are wet snow and large, wet hail. This makes it a valuable product for severe weather forecasting, particularly when evaluating a storm over time.

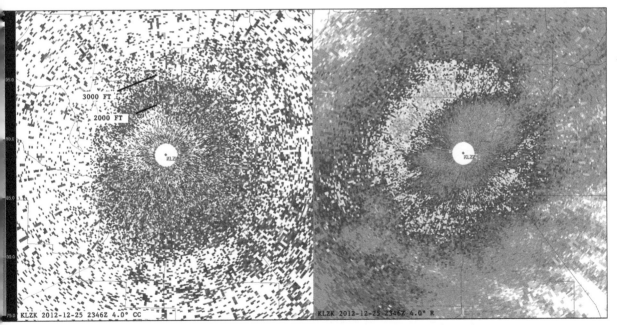

Figure 5-3. Melting level as seen with correlation coefficient (left) and standard reflectivity (right). The melting level is at the top of the bright band, so in this case it corresponds to a height of 3000 ft. At the time, stations at the surface were receiving rain, but this transitioned to snow about 2 hours later as cold air advection infiltrated the area. Example is from 25 December 2012, 2346 UTC, at Little Rock AR.

5.4. Differential propagation phase (ϕ_{dp})

Differential propagation phase (simply ϕ_{dp}) is defined as ϕ_{hh} - ϕ_{vv} and is expressed in degrees of phase shift. So it is an expression of difference between the co-polar horizontal and vertical phases. The phase shift occurs in weather phenomena and does not "revert" to any sort of base value with increasing range, so images of differential phase will often be highly streaked along many of the radials. Because of this, differential phase is rarely used by operational forecasters. Instead, specific differential phase shift (K_{DP}) is used, which emphasizes where the differential phase changes occur and shows values of zero where no phase change was measured.

5.5. Specific differential phase (K_{dp}, KDP)

Specific differential phase or differential phase shift is abbreviated formally as K_{dp} and informally as KDP. It is the range derivative of differential phase and expresses how much the raw differential phase changes over a given distance. In other words, the *greater the change in differential phase* from bin to bin, the higher K_{dp} will be. In essence, it highlights gradients of differential phase rather than the actual values themselves.

Since gradients are a function of the sampling size, which may be noisy at small distances and oversmoothed at long distances, there are not "typical" sets of K_{dp} values which can be compared across all polarimetric radars. However since the WSR-88D radars use the same hardware, the same gate spacing, and the same algorithms, the K_{dp} values on WSR-88D radars can be compared between radars. Overall it is more important to recognize high and low values of K_{dp} than to look for specific ranges of numbers.

The K_{dp} parameter can be used to identify areas of highly non-spherical scatterers, such as raindrops, which have an oblate shape as they fall. Since this tends to filter out hailstones, K_{dp} tends to be sensitive to rain only, unlike reflectivity which has an affinity for large, wet hailstones. The K_{dp} parameter also filters out ice crystals and snow. So in winter situations it can help single out areas of rain.

This product can also be used as a proxy for reflectivity. It may be questioned why this should be done, since perfectly good values of reflectivity are usually available and K_{dp} returns are often weak. Phase shifts are immune to beam blockage, so if a radial cuts through widespread or intense precipitation or grazes mountainous terrain, K_{dp} will still provide more accurate estimates of intensity than reflectivity. So not only does heavy rain return power, it also produces a phase shift, and either of these can be used to detect it in lieu of the other.

Another emerging use of this is to identify areas of thunderstorms, where ice crystals orientations are responding at a small scale to changes in the electrical field within a thunderstorm. The K_{dp} value has a tendency to increase sharply before an in-cloud lightning flash occurs, then diminish back to its original

---

KDP explained

Specific differential phase is:

0.5 ($\Delta\phi_{dp} / \Delta r$)

which shows that it equals the change in raw differential phase over a given length of radial distance. From gate to gate, radial distance remains constant, so KDP simply provides the change in raw differential phase from gate to gate.

KDP rule of thumb

As a general rule of thumb, heavy rain is signified by a KDP phase shift of 1° per kilometer.

LDR: What is high? What is low?

Shown here is an approximate listing of linear depolarization ratio values:

< -30	Low
-29 to -25	Medium
> -25	High

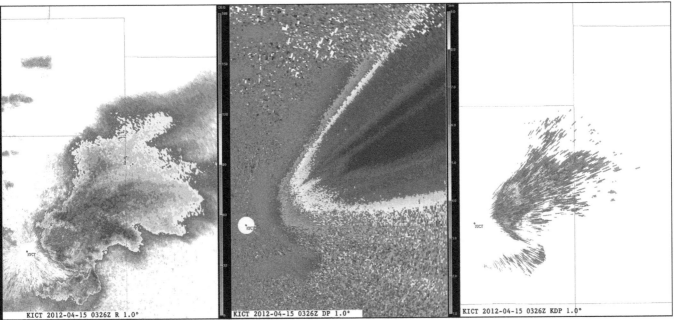

value in a matter of minutes. This technique is probably more effective for research radars with high scan rates, but it is possible that lightning activity could be sensed by dual-polarization WSR-88D radars in this manner.

5.6. Linear depolarization ratio (LDR)

The linear depolarization ratio (LDR) is a measure of the ratio between cross-polar power and co-polar power, in other words Z_{HV} / Z_{HH}. It senses where precipitation or other scatterers are depolarizing the radiation from the antenna. Particles that are tumbling, wobbling, or have complex shapes are more likely to have elevated LDR. Much like the correlation coefficient, this gives information on precipitation diversity. This product is not offered with polarimetric products as much as the other quantities, so this section will only discuss it briefly.

5.7. Polarimetric algorithms and derived products

Though the RPG does produce a few derived products from the base polarimetric data, these will be described here rather than in the Derived Products chapter to help provide the material within its proper context.

5.7.1. MELTING LAYER DETECTION ALGORITHM (MLDA). The MLDA is a critical part of allowing the RPG to correctly identify precipitation type. This algorithm finds the "bright band", or melting level, and tries to identify its height and dimensions. To do this, it looks at specific reflectivity, correlation

Figure 5-4. Differential phase products as compared with the standard base reflectivity product. This is an HP supercell in southern Kansas on the evening of 14 April 2012. Note that the radar site is in the lower left of each frame. The leftmost frame shows standard reflectivity and the classic appearance of an HP supercell. The center frame shows raw differential phase shift. Echoes always have a streaked appearance since the phase changes as the wave propagates through the precipitation, then remains at that phase unless other scatterers are encountered further out. This raw product is not used by operational forecasters. Rather, specific differential phase shift, or KDP (right) is preferred, which is simply the gradient of the phase shift. This product is useful. While it is not sensitive as base reflectivity, echoes further away cannot be attenuated by nearby precipitation.

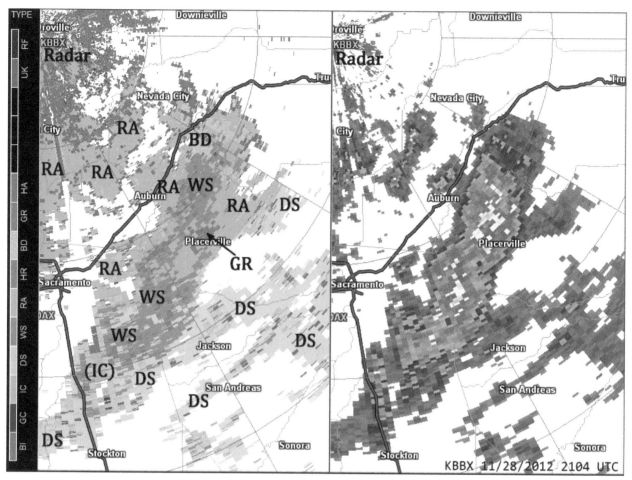

Figure 5-5. An example of the Hydrometeor Classification Algorithm (left) viewed in GRLevel3 and compared with reflectivity (right) on the 0.5° tilt. For clarity the dominant HCA areas have been labeled with a bold typeface. The location of the radar site at the top left and awareness of the conical nature of the tilt are important for proper interpretation of this image. In this context, rain (RA) and big drops (BD) can be seen at the lowest levels, with wet snow (WS) at slightly higher levels, transitioning to dry snow (DS) at the highest levels. Note the radial artifact extending through the entire image which intersects the area marked (IC). This is the result of beam blockage, which degrades the HCA result and attenuates radar reflectivity.

coefficients, and signal-to-noise ratios associated with a bright band and assumes it consists largely of wet snow.

The RPG also offers a melting layer (ML) product which plots the normal correlation coefficient product and overlays two solid rings corresponding to the height where the radar beam intersects the bottom and top of the melting layer. Outside of these rings, there are two dashed rings which indicate where the radar detects the edges of the melting layer. Anything outside these four rings is either entirely above or below the melting layer.

5.7.2. HYDROMETEOR CLASSIFICATION ALGORITHM (HCA). The HCA is an algorithm that does a lot of the dirty work in interpreting differential reflectivity, correlation coefficient, and other polarimetric base products to specifically identify the dominant type of precipitation within each gate. It must be emphasized that a lot of the work on these algorithms are still in their early stages and misclassifications are likely to occur. Forecasters need to be acquainted with all the polarimetric base products, not only to check

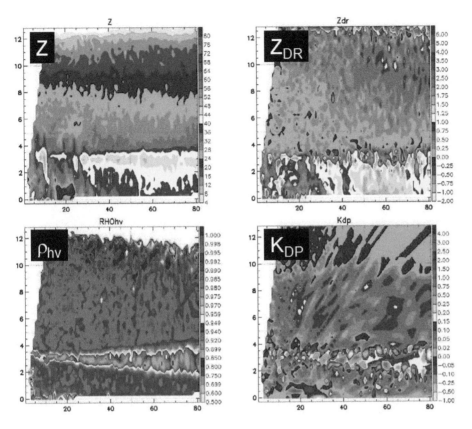

Figure 5-6. The melting level as observed in cross-section form. Clockwise from top left are base reflectivity, differential reflectivity, specific differential phase shift, and correlation coefficient. It can be seen that the correlation coefficient (ρ_{hv}) shows the bright band with great precision. Also note that the differential reflectivity (Z_{DR}) is well above 1 below the melting level, signifying rain, but is 0 to 1 above the melting level, suggesting ice crystals. *(From Kevin Scharfenberg, CIMMS/NSSL, "Future QPE: Dual-Pol and Gap-Filler Radars", 2007)*

Ice crystal habitats

Typical habitats for ice crystals are as follows:
Needles   -4 to -6°C
Columns   -6 to -10°C
Plates   -10 to -22°C
Dendrites -12 to -16°C

for consistency of the HCA product but to gain more insight into the weather situation and improve experience with the polarimetric parameters.

The HCA divides precipitation into twelve categories, listed in the sidebar. It uses all of the polarimetric products, plus the melting layer detection algorithm, along with fuzzy logic to try to properly classify each bin. Fuzzy logic is necessary, since the relationship between different types of precipitation on polarimetric products is somewhat tenuous and a flowchart or decision-tree type approach will likely fail. The most critical piece of information that enters the HCA algorithm is the melting layer height. This helps the algorithm discard unlikely situations such as wet snow above the melting layer or small ice crystals beneath. There are other rule sets that are part of the algorithm. For example an identification of hail will be rejected if the reflectivity is below 30 dBZ, and an identification of biological sources will be rejected if the correlation coefficient is less than 0.97.

One noteworthy problem that affects the HCA algorithm is a warm-over-cold layer, as might be found north of a warm front. The algorithm handles the melting layer, but not a *refreezing layer* below if a cold air mass exists beneath. Cold rain, freezing rain, or ice pellets may occur at the surface regardless of what the HCA indicates. The forecaster must synthesize the radar output with a thorough understanding of temperature profiles in the lower troposphere across the forecast area.

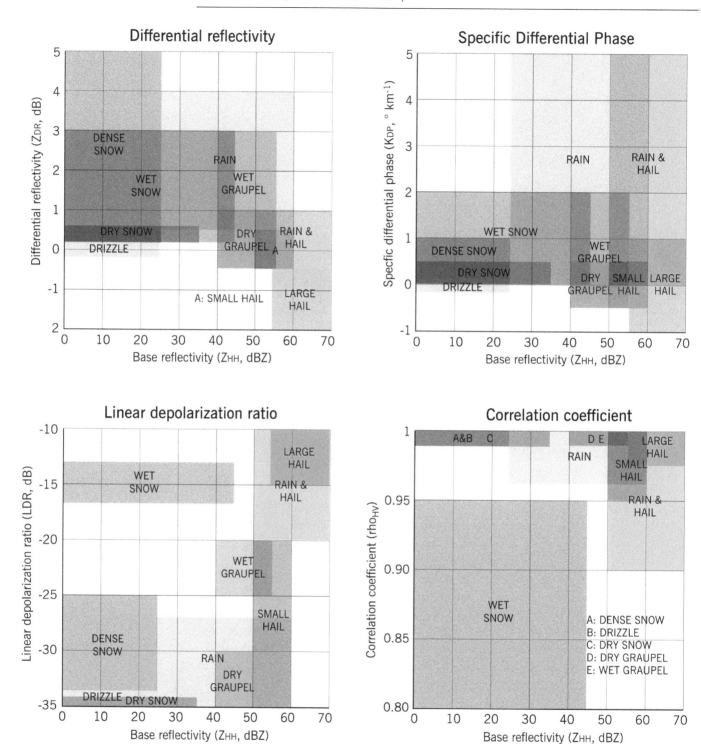

Figure 5-7. Summary of typical polarimetric parameters for various types of precipitation with respect to base reflectivity. The K_{DP} values here are specific to the WSR-88D. The data for this table is adapted from the work of Bringi and Chandrasekar (2001), in turn adopted from Straka and Zrnic (1993).

With this in mind, it must be remembered that the HCA was developed primary for warm precipitation such as rain, showers, and thunderstorms. While it is gradually being adapted to handle winter weather situations, failures are still possible and forecasters must anticipate conditions in which the HCA will fail, and even in ideal conditions the HCA should be crosschecked with base polarimetric products and surface observations. The HCA and MLDA will likely be improved and refined in upcoming WSR-88D builds.

5.8. Differential diagnosis

Differential diagnosis techniques can be used with polarimetric radar to identify which of two likely precipitation types might be actually occurring.

5.8.1. RAIN VS. SNOW. To discriminate between rain and snow, differential reflectivity (Z_{DR}) is the best parameter because it separates the oblate, high Z_{DR} rain drops from prolate, low Z_{DR} ice crystals. Dry snow has a Z_{DR} averaging 0 to 1, wet snow averages 1 to 2, and rain averages 2 to 6. Differentiation is more difficult in weak precipitation regimes, since drizzle is made up of spherical particles whose Z_{DR} averages about 1 to 2, within the same range as wet snow. Also the K_{dp} parameter can discriminate between rain and snow.

5.8.2. RAIN VS. ICE PELLETS. As with rain versus snow, the best discriminator for ice pellets (sleet) is differential reflectivity, Z_{DR}. Ice pellets tend to be somewhat prolate and will have a Z_{DR} of 1 or less, while rain has a Z_{DR} of well above 2. Differentiation becomes difficult when ice pellets are melting or drizzle is the primary type of liquid precipitation, since both of these have spherical or slightly oblate signatures and will have a Z_{DR} of 1 to 2.

5.8.3. RAIN VS. HAIL. Hail particles have a Z_{DR} of less than 1 while rain has a Z_{DR} exceeding 2. However in a mixed rain/hail situation Z_{DR} becomes less reliable, so the LDR should be used to identify zones of mixed liquid and ice. Rain or very small hail occurring alone without a precipitation mix is associated with LDRs of about -30 to -40, while mixed types are associated with LDRs of -20 or higher.

5.8.4. SNOW VS. ICE PELLETS. Ice pellets (sleet) shares low Z_{DR} characteristics with snow. However ice pellets show higher reflectivities due to the larger particle size, especially if they are melting. Ice pellets usually return a slightly higher LDR, typically -25, while snow usually remains below -35.

Basic HCA categories
Listed here are the basic hydrometeor types given in the WSR-88D classification scheme.

DS Dry snow
WS Wet snow
IC Ice crystals
BD Big drops
RA Rain (light and moderate)
HR Heavy rain
GR Graupel
HA Hail (mixed with rain)
UK Unknown
GC Ground clutter / AP
BI Biological
ND No data (less than threshold)
Coming soon: tornado debris

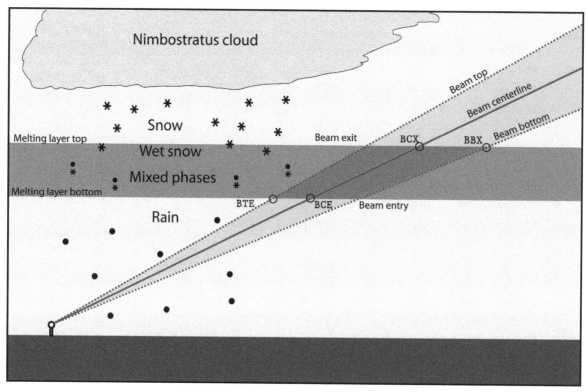

Figure 5-7. Melting level and its relation to the radar beam. There are four levels of importance to forecasters. First is where the centerline of radar energy enters and exits the melting layer. This is represented by the beam center entry (BCE) and beam center exit (BCX) levels. The other two heights of interest are where the illuminated volume enters and exits the melting layer. This is bounded by the beam top entry (BTE) and beam bottom exit (BBX). It can be seen that from near to far, the sequence is BTE-BCE-BCX-BBX. Many NEXRAD products, journal articles, and forecast reviews show these four range rings without labeling, so understanding this progression helps make sense of which ring is which.

5.8.5. SNOW VS. HAIL. Use LDR. Snow is associated with LDRs of about -30 to -40 while hail is associated with LDRs of -20 or higher, particularly if large and/or wet.

5.8.6. ICE PELLETS VS. HAIL. Structurally ice pellets are similar to hail, but are smaller. However LDR is a useful discriminator, with ice pellets having an LDR of less than about -25 and hail exceeding -25 to to the wider diversity of particle types. Better clues may be found by assessing the meteorological pattern and determining whether instability can support deep moist convection and hail.

5.9. Features and phenomena

5.9.1. MELTING LEVEL. The melting level has been recognized as an important feature on conventional radar for over 50 years. It takes on special significance in polarimetric radars because many additional parameters about it can be extracted. For example, in a typical situation, ascending with height we would see reflectivity (Z) increase within the melting level, peaking at the bottom of the layer where wet snow (water-coated ice) dominates, and from there it decreases again. If we look at the differential reflectivity (Z_{DR}) value, we would see it increase with height at the bottom of the wet snow layer, then fall off

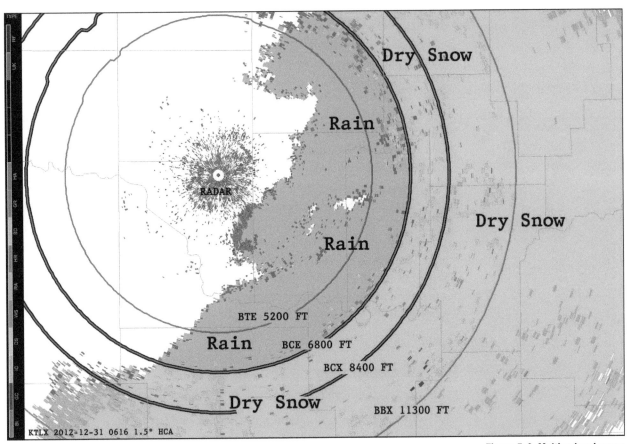

Figure 5-8. Melting level range rings on an HCA (hydrometeor classification algorithm) product on the back side of a cold rain event. The precipitation types suggested by the HCA product are printed in bold. The four basic melting level range rings are shown here. It can be seen that the HCA switches the precipitation type from rain to dry snow at the BCE (beam center entry) height. However, the melting level is being sampled most strongly between the two thick lines, and to a lesser extent between all four lines. The volumes between the two thick lines (BCE and BCX) are likely to be a mixed phase precipitation regime, not dry snow.

again above the melting layer as ice crystals predominate. Using the correlation coefficient (ρ_{HV}), which is the best indicator of the melting layer, we would see it drop off sharply as it enters the melting layer, reaching a minima within the layer containing mixed snow and water (just below Z and Z_{DR}), then increase again.

5.9.2. TORNADO DEBRIS. Although tornado debris often shows a distinct area of high intensity and high spectrum width near the tip of a hook echo, it may also show low differential reflectivity and low correlation coefficient. Tornado debris is expected to be added to the HCA algorithm soon.

5.9.3. DEPOLARIZATION. Lightning strikes within a thunderstorm produce significant electromagnetic fields which can briefly change the orientation of ice crystals. This may produce brief, localized changes in the polarimetry of the upper part of the thunderstorm where ice crystals predominate. The change usually lasts on the order of seconds but may be detected within a radar sweep, particularly at a single tilt. It may appear on the differential reflectivity product, correlation coefficient, and KDP product as radial spikes within the ice crystal region which change from scan to scan or which quickly disappear.

5.10. Filtering

By selectively identifying patterns associated with precipitation, polarimetric radar display packages have the ability to filter non-hydrometeors from reflectivity and velocity products. One of the most important discriminators is ρ_{HV} (the cross-correlation coefficient) as described earlier. This does carry the problems of masking important features such as fronts and pre-convective boundaries, so if filtering is employed in a radar display software package, the forecaster usually has some ability to raise or lower the thresholds so that these features are not removed.

This technique does not eliminate the problems with second-trip echoes, since those are legitimate precipitation features that will carry all of the correct signatures for hydrometeors and will pass through the filter.

Overall, this type of filtering is only done in research radar work and in professional-grade radar display packages and workstations, so it is currently only of minor interest to operational forecasters. However as this type of filtering is further refined and becomes employed in third-party WSR-88D display software, at the very least, it is a technique that deserves mention.

REVIEW QUESTIONS

1. In a vertically polarized wave, in which direction are the electrical and magnetic fields oscillating?

2. What does the term Z_{HV} mean?

3. If Cloudy With A Chance of Meatballs came to life and hot dogs began raining out of the sky, remaining horizontal (its ends side to side, not up and down), what kind of values would differential reflectivity show? Why is this? The differential reflectivity formula is $10\log_{10}(Z_H/Z_V)$?

4. What shape does a large raindrop have? Is it wider in the vertical or horizontal, and what type of effect does this have on differential reflectivity?

5. If the vertically polarized power return is the same as the horizontal power return, what value is Z_{DR}?

6. In general, does hail have an oblate, prolate, or spherical signature on radar?

7. Because of the oblate shape of birds within a large flock and the uniformity of the individual oblate shapes, what polarimetric characteristics will this flock have?

8. What do low values of CC (below 0.90) signify? What are some possible exceptions to this rule?

9. How does specific differential phase (KDP) relate to its parent product, differential propagation phase (theta$_{DP}$)?

10. Which polarimetric product can be used to find heavy rain in a situation in which attenuation by nearby thunderstorms might degrade reflectivity?

```c
            max_shr = L->data.shear;

if (L->data.maxgtgvd > gtgmax)
   gtgmax = L->data.maxgtgvd;

/* update L and last_node */
                        last_node = L;
L = L->next;

      } /* END of while (L->next != NULL) */
   } /* END of else of "if (YES_DO_SMOOTH)" */

   /* calculate the center of the feature: mid-point between
    * the average maxmimum and minumum velocity location.
    * Also calculate the azimuth and range of the locations
    * of the locations of the velocity minimum and maximum,
    * the radial diameter.
    */
   azmin = vmin_ptr->data.beg_azm;
   azmax = vmax_ptr->data.end_azm;
   rmin = vmin_ptr->data.range;
   rmax = vmax_ptr->data.range;
   x1 = rmin * sin(azmin * DTR);
   y1 = rmin * cos(azmin * DTR);
   x2 = rmax * sin(azmax * DTR);
   y2 = rmax * cos(azmax * DTR);
               dia = sqrt((x1 - x2) * (x1 - x2) + (y1 - y2) * (y1 -
   cx = 0.5 * (x1 + x2);
   cy = 0.5 * (y1 + y2);
   cr = sqrt(cx * cx + cy * cy);
   ca = atan2(cx, cy) / DTR;
   if ( ca < 0.0 )
ca = DEG_CIR +ca;

   /* calculate range/azimuthal distance aspect ratio */
                  hr = fea
   lr = last_node->data
   ratio = (hr - lr) /

   /* calculate the fe
   ht = cr * sin(ele

   /* Check to see if
    * Delete features
```

6 DERIVED PRODUCTS

6 | Derived Products

Base products measure the basic properties of backscattered energy returned from weather systems and other processes occurring near the radar. The values are provided as-is, with corrections and enhancements based mostly on principles of electromagnetic engineering. On the other hand, derived products refer to a class of images and data that are produced with base images and which transform the data in some way based on *meteorological* relationships. They are products which are very much the invention of meteorologists specializing in radar technology and are designed for warning and forecasting purposes.

Unfortunately there is no perfect algorithm: for every derived product that has been created, there are inherent problems in the measurement or uncertainties in the solution. Even the simplest derived product, composite reflectivity, is not consistent between near and distant ranges because vastly different altitudes are being sampled and beam broadening yields very different resolutions at different ranges. Because of these drawbacks and the absence of an artificial intelligence mechanism to help overcome these deficiencies, a warning or forecast should never be issued solely on any derived product. The forecaster must still consult the base products, check for consistency, diagnose the weather phenomena, and then make decisions based on this information.

Derived products and their underlying algorithms tend to be specific to radar equipment and radar networks. Since the only set of derived products that are widely available to the general public are those offered by the United States WSR-88D network, this chapter will focus exclusively on those derived products. Since the 1980s the WSR-88D algorithms have had a long history of testing, operational use, and refinement, so all the content presented here will be valuable to those in other countries whose weather services have other types of radar equipment, even if the algorithms are not exactly the same as those used by the American network.

6.1. Composite reflectivity

The composite reflectivity product, often abbreviated CR or CZ, is one of the simplest derived products. Using all radar bins in a vertical column, it shows the the highest detected reflectivity value, regardless of tilt. This is useful for general surveillance in a fair weather pattern as well as for monitoring a situation before precipitation develops, when the droplets or ice crystals may form at a much higher level than that shown by the standard 0.5° tilt. It is also useful for elevated convection and high-based precipitation, when the 0.5° tilt might be in the lower portion of the virga or well below it. In the latter case, the radar will probably be overestimating precipitation at the surface, but the forecaster will still be able to monitor precipitation areas and convective cells occurring aloft.

This product, since it uses vertical stacking of bins, will present a distorted picture of the reflectivity patterns when significant vertical wind shear exists. The difference between low-level and high-level winds will cause updrafts and downdrafts to tilt rather than follow a vertical trajectory.

Adaptable parameters
Nearly all WSR-88D derived products provide weather service offices with sets of adaptable parameters. This changes various thresholds which in turn affects the quality and results of the derived product. For example, the forecaster can change the coefficient of the rain rate equation so that the precipitation totals products are more representative. Therefore minor differences may sometimes be seen when comparing products between radars and even between different dates. The adaptable parameter values are transmitted as part of the Level III data stream and can be reviewed with radar viewers or opened in text programs.

Title image
Actual WSR-88D source code for the mesocyclone detection algorithm. In this portion shear is being calculated. A total of 132,000 lines of FORTRAN code and 910,000 lines of C++ code currently make up the WSR-88D software build. All of this creates the products that have become the backbone of mesoscale meteorology in the United States.

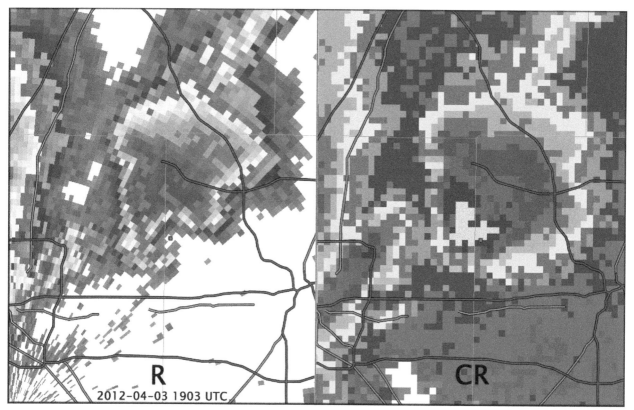

Figure 6-1. Comparison of base reflectivity and composite reflectivity products, showing the Arlington-Irving, Texas supercell of 3 April 2012 at the same instant. The reflectivity image is for the 0.5° tilt, and shows a very distinct hook echo on the southwest side. The composite reflectivity image shows almost all of these features smeared out, with the highest reflectivities dominated by large hail at higher elevations. This illustrates why composite reflectivity should not be used for storm-scale analysis, and why forecasters need to ensure they are using a base reflectivity product when using an unfamiliar radar source. The approximate position of the storm's low-level updraft is shown by the small dot on the southern part of the storm.

The composite reflectivity product should never be used for identifying boundaries, tornadic wind circulations, and low level features, because all of these may be masked by stronger reflectivities at higher elevations. Even if a feature looks like a hook, a supercell, or a fine line on composite reflectivity, the individual gates may be built up of reflectivity maximums from a variety of different tilts.

6.2. Vertically integrated liquid

The vertically integrated liquid (VIL) product shares a lot of commonality with composite reflectivity in that it evaluates reflectivity vertically above a given point. A VIL value is highest where intense reflectivity exists above a given point at many different tilts. Instead of finding the highest value, it integrates them vertically to give an estimation of the total liquid above a given point. It is given in units of kg m^{-2}.

The same issues with composite reflectivity will affect VIL. But while composite reflectivity is just a measurement of maximum reflectivity in the vertical, VIL indicates a "stacking" of high intensity in the vertical with the assumption that all of this precipitation will fall downward through the column.

If environmental shear is present, it degrades the meaning of the VIL estimation because it is no longer likely that precipitation will fall downward through the column. This is one of the greatest problems with the product.

One advantage of VIL is that it can be used in conjunction with reflectivity to help weak features like fronts, snow bands, and smoke plumes stand out more strongly from ground clutter. This is because reflectivities are integrated with height, and if the feature is present at a given point at two tilts the features will combine additively while ground clutter remains at its original intensity or gets canceled out.

With the WSR-88D there are two expressions of VIL: legacy VIL and digital VIL (DVIL). While legacy VIL uses a cartesian grid, the DVIL product is in a radial format and does not use truncation. Furthermore legacy VIL is "truncated" at 56 dBZ, which means that reflectivity at any given level exceeding 56 dBZ will be treated as 56 dBZ to minimize contamination of the VIL value by hail. The DVIL product does not use this truncation, and this can make the result significantly different when hail is present.

6.3. Echo tops

The echo tops (ET) product examines all of the tilts above a given location to find the highest elevation at which a reflectivity of 18.5 dBZ or more is detected (a value which can be modified by the radar operator). This constitutes the assumed location of precipitation echo tops. It heights that are slightly higher than any "cell tops" identified by the storm tracking algorithm, which use 30 rather than 18.5 dBZ as the threshold.

The WSR-88D radar only has limited resolution in the vertical. This is because at a moderate to long range, and depending on the VCP, only a few tilts may be available at the cloud tops and the estimated cloud top height is a function of range and which of these few available tilts sampled a particular cloud top. If a 45,000 ft storm top is sampled by a beam that only intersects its upper part at 35,000 ft, 55,000 ft, and 75,000 ft, it will be measured as having a height of 35,000 ft.

As an effect, where a broad, stratified area of precipitation exists, such as in the upper parts of a thunderstorm complex and within stratiform precipitation areas, the echo top product will show a "stadium effect" with a jagged, concentric pattern centered on the RDA caused by the scarcity of available tilts at cloud top height. Looking at the stadium effect along a given radial, echo tops rise gradually with increasing distance until the beam leaves the cloud top, then sharply descends to the next tilt, rises gradually with that beam, sharply descends to the next tilt, and so on. So the ability of the radar to accurately measure the echo tops is limited, and dependence on exact values should be avoided. Sidelobes can further distort the echo top accuracy.

A new product called enhanced echo tops (EET) was introduced to WSR-88D builds in 2010. This algorithm interpolates across the "stadium

The VIL equation explained
VIL is a vertical integration of the formula:
$$M = 3.44 \times 10^{-3} Z^{4/7}$$
where Z is the radar reflectivity factor and M is the liquid water content in g m^{-3}.

Rain rate adjustment
The precipitation processing subsystem of the WSR-88D contains an adjustment algorithm which, if enabled by the site operator, allows it to ingest a table from the AWIPS system called Multi-Sensor Precipitation Estimator, which compares rain gauge data to radar estimates and produces a bias table.

The rain rate equation
The relationship of precipitation to reflectivity as used by the WSR-88D is $Z=300R^{1.4}$ by default, where R is the rainfall in mm h^{-1} and Z is the reflectivity factor in dB. However this is one possible approximation; five of the most well known ones include:

$Z=300R^{1.4}$ Convective
$Z=250R^{1.2}$ Tropical
$Z=200R^{1.6}$ Stratiform
$Z=130R^{2.0}$ East U.S. (cool)
$Z=75R^{2.0}$ West U.S. (cool)

effect" levels, reducing the concentric artifacts and smoothing the resulting output. The algorithm also recognizes where cloud tops are ascending into the cone of silence, and paints these areas with a color code labeled as "topped". This means that the radar was unable to determine a cloud top height.

6.4. Precipitation estimation

The WSR-88D has brought enormous advances to the science of flash flood forecasting. One of the driving forces to develop the manually digitized radar grid in the 1970s was not the ability to have detailed national radar composites but to feed the data to early microcomputers and allow them to make numerical estimates of total precipitation. Through the 1980s flash flood forecasting mainly relied on this type of guidance but still required a lot of manual radar interpretation and subjective techniques to identify areas of concern. The WSR-88D's arrival in the mid-1990s brought a quantum leap in the ability to identify areas of heavy precipitation down to the resolution of 1 nm, and by the early 2000s sophisticated precipitation estimate systems that blended historical radar data and observed totals were in use at hydrometeorological prediction

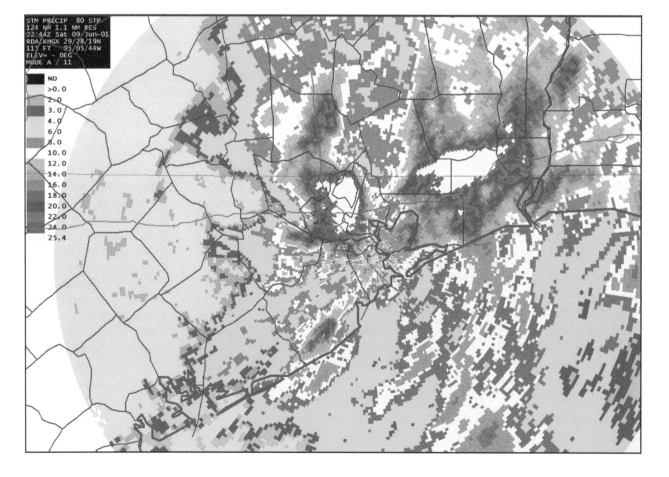

Figure 6-2. Storm total precipitation for tropical storm Allison as of 2344 UTC 9 June 2001 in southeast Texas. This clearly identifies the hardest-hit areas of Houston, at center, and Beaumont, to the center east. By the time the precipitation had finished, some areas had measured over 40 inches of rain. This graphic was created by the Digital Atmosphere weather analysis package.

centers across the country and formed the framework for numerical modeling of flash flood danger.

The WSR-88D RPG runs a precipitation processing subsystem (PPS), which is an algorithm that generates all of the precipitation totals. All of this is based on the reflectivity in the lowest tilt (usually 0.5°) that is not blocked or contaminated with ground clutter. The algorithm is based upon the simple relationship of reflectivity to liquid water, but it also takes into account bias from different VCPs, ground clutter filters, and manual adjustment values. From this a precipitation rate is determined, and the time between volume scans yields the precipitation since the last volume scan.

In terms of products, the RPG offers one-hour precipitation (OHP), in other words precipitation over the past 60 minutes; three-hour precipitation (THP), which is the precipitation over the past 180 minutes; and storm total precipitation (STP), which is the precipitation over a number of hours or days. Since high precipitation amounts spread out over a longer duration has less

A word of warning
Never issue a warning based solely on output from derived products. The algorithms are not infallible, and they don't replace proper minute-to-minute diagnosis of the storm situation using base products in conjunction with surface, upper air, satellite, and even visual data.

Figure 6-3. Tornado Detection Algorithm (TDA), shown by the downward-pointing triangoe, and Mesocyclone Detection Algorithm (MDA) output, indicated by a yellow circle, during a tornadic event in Irving, Texas. Both the TDA and MDA output are computed separately, and both show good agreement in this example.

WSR-88D precip abbreviations
One-hour   OHP
Three-hour THP
Storm total STP

The TVS
Detection of a TVS by the WSR-88D TDA algorithm requires the following:
- Shear correlated vertically ("3-D").
- A circulation depth of 1.5 km or more with a base of less than 600 m above the surface or at the 0.5° tilt.
- Gate-to-gate shear of at least 70 kt somewhere in the circulation or of at least 49 kt at the base of the circulation.
- Shear values meeting the adaptable parameter thresholds customized at each individual radar site.
- Values displayed in the attribute tables will include low-level shear (LLDV), maximum shear (MDV), average shear (AVGDV), a base altitude of the circulation (BASE), and circulation depth (DPTH).

of an impact, the forecaster should always check the storm total precipitation product to determine the approximate number of hours it covers. The storm total precipitation is reset by the RPG whenever reflectivity intensity and area falls below a customizable threshold (20 dBZ covering 80 km^2, by default), and can also be manually reset by the radar operator. In a rainy weather pattern, the storm total precipitation may represent an accumulation of many days worth of data.

The default precipitation total algorithm is based on one specific approximation of rainfall rate to reflectivity factor ($Z=300R^{1.4}$), and this is designed for convective weather regimes. As a result, the algorithm will tend to underestimate warm rain event, in which precipitation development areas are below the freezing level. On the other hand, it tends to underestimate precipitation in cold weather regimes. The WSR-88D RPG does allow operators to change the parameters of the rain rate equation and to cap the overall rainfall rate to filter out spiking of the reflectivity by hail.

When any part of the precipitation column consists of ice or mixed phases, the approximation does not work well because it is based on Rayleigh scattering by water droplets. In a typical thunderstorm, a large part of the precipitation area does in fact extend well above the freezing level past supercooled water regimes. While ice crystals and graupel in the upper part of the storm do affect the precipitation estimate, they only contribute a small amount of degradation. Much more significant degradation is caused by hailstones at any level within the thunderstorm. When hail occurs, the power returned to the radar is much higher and the precipitation algorithms will overestimate rainfall.

The next problem that affects the precipitation algorithm is when the precipitation column is tilted. The radar assumes that all precipitation falls directly downward. A tilted precipitation column will be caused by any environment with strong deep-layer shear. This not only impacts the rainfall estimates but may cause them to be misplaced by up to several miles.

One further issue is in fast environmental flow, when precipitation systems are moving rapidly through the radar volume. Consider a squall line moving 60 mph through a volume scan being sampled once every 5 minutes. This means that the squall line at a given tilt will appear to "jump" forward 5 miles with each successive scan. Since this effectively tilts the precipitation column, it will introduce errors into the precipitation algorithm. Fortunately, it can usually be identified by the distinct herringbone or buzzsaw appearance in precipitation algorithm patterns.

Polarimetric radar promises to greatly improve the accuracy of the WSR-88D PPS since it can identify precipitation type within each gate and use this data to approximate rainfall rate and storm total. Due to the longer chain of assumptions that go into the final product, the quantities it produces tends to be noisy, and artifacts may be seen along features such as the melting level top and bottom. However these algorithms are still in development and as they mature and are perfected they are likely to replace the legacy precipitation processing subsystem in the years ahead.

6.5. Storm detection algorithm

The WSR-88D radar product generator features an algorithm for identifying and tracking storms called the Storm Cell Identification and Tracking (SCIT) algorithm. This has long been the backbone of many of the forecast products offered by the radar. It is important to emphasize it is based entirely on reflectivity, and does not consider velocity products nor output from the hail, tornado, or mesocyclone algorithms.

The algorithm starts with the base reflectivity product at each tilt and looks along radials for one-dimensional *segments* of high reflectivity. These segments are then compared across radials to form two-dimensional storm *components*. These centroids are then associated vertically through different tilts to form three-dimensional features known as *centroids*. The centroid is located roughly at the cell's three-dimensional center of mass. There are various thresholds specifying properties like minimum reflectivity, segment length, and so on that the segments, components, and centroids must meet, and NWS offices may adjust these further in the WSR-88D's adaptable parameters settings.

Figure 6-4. Mesocyclone (circle), hail (upright triangle), and tornado (inverted triangle) near Fort Wayne, Indiana during the 29 June 2012 derecho. The tornado icon shows good correlation with the mesocyclone location and corresponds to a notch on the forward side of the squall line.

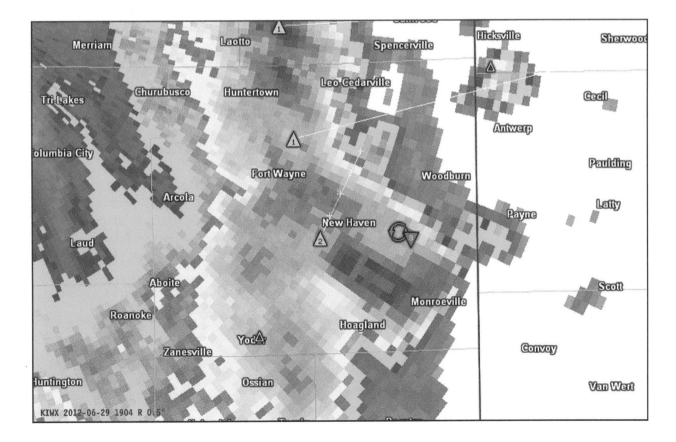

The storm detection algorithm not only locates and identifies storms but calculates basic properties associated with it. This is supplied with each volume scan as a storm attributes table. One of the fields in the table is a two-digit code identifying each storm on both graphics and text output, such as "D0", "E1", etc. In general, the lower the number, the older the cell.

Another important field is the storm position forecast (storm track) information. This may be overlaid graphically as a projected track of the storm centroid. It is calculated based on current and past centroids, and it must be remembered that this does not correspond to the track of mesocyclones and tornadoes but rather the location of the storm's core. For new cells it is based on the average movement of all cells or from environmental information, and once a history is established the storm forecast is based on a history of the storm's movement. If errors are found for previous forecasts of storm track, the projected track is shorted from 60 minutes down to 45, 30, or 15 minutes depending on the magnitude of the error. So truncated storm tracks reflect uncertainty for that cell in the algorithm.

Other fields found in the attributes table are the maximum intensity (dBZ) found by the storm identification algorithm, the maximum echo top, and output from the mesocyclone and tornado algorithms. These are provided as a text table which may be viewed in a separate window, but some elements may be plotted on the radar graphics depending on the display software.

6.6. Mesocyclone detection algorithms

The Mesocyclone Detection Algorithm (MDA) is an algorithm used in the WSR-88D to detect rotation on the order of 1 to 10 km. First it detects "2-D" features by looking for horizontal shear, then ranks it according to its intensity. It then associates 2-D features from different tilts to produce a 3-D centroid. Then, these centroids are related from scan to scan, in other words, with respect to time, to produce a 4-D detection. If horizontal (2-D) shear patterns cannot be built up vertically (3-D), then it is considered to be uncorrelated shear (UNCO), and if the resulting 3-D centroid does not meet the threshold values of width and length, it is considered to be 3-D correlated shear (3DCO). Otherwise it is classified as a mesocyclone. This 4-D detection information is disseminated along with extrapolated tracks and can be overlaid on products.

Symbols used for MDA mesocyclone locations consists of a circle. With GRLevelX it is shown as a broken circle for elevated mesocyclones and continuous for non-elevated ones. Mesocyclones are further ranked according to strength, with green colors used for weak ones and red to purple ones for strong mesocyclones.

Forecast tip
* Some forecast offices use a POSH value of 70% as a threshold value for issuing a warning.
* The maximum reported hail size is usually less than that generated by the MESH value.
* Spotter reports are essential for confirming hail sizes.

6.7. Tornado detection algorithms

The Tornado Detection Algorithm (TDA) is designed to detect rotation on with a scale of less than 1 km. The signature of a tornado is known as the *tornadic vortex signature* (TVS). The tornado by definition is a violently rotating column of air in a thunderstorm, and is on a smaller scale than the mesocyclone. Accordingly, the algorithm is designed to detect rotation specifically of less than 1 km. Effective with Build 10 in 2008, the TDA algorithm replaced the older TVS algorithm, whose performance was considered to be mediocre at times.

The TDA works in much the same way as the MDA, by associating 2-D shear into 3-D centroids, and then with respect to time to form a 4-D TVS detection. The difference is that the 2-D search algorithm looks specifically for gate-to-gate shear, and the detection thresholds allow for a wide range of scales and strengths so that everything from wedge tornadoes to landspouts can be detected. The depth of the TVS must be 1.5 km or more with its base at less than 600 meters above the surface or at the 0.5° tilt. If the base is above 600 m and above the 0.5° tilt, it is considered an ETVS (elevated TVS).

The standard symbol used on most software systems for a tornadic vortex is a downward-pointing triangle, with hollow symbols reserved for a elevated TVS. With GRLevelX, a number is plotted in the center representing tens of knots of low-level shear (LLDV). Forecasters can use the TDA to confirm the likelihood of a tornadic circulation if a TVS is identified in the right rear quadrant of a severe thunderstorm or along a notch or appendage. Since the algorithm is completely independent from the MDA algorithm, a TVS identified within a mesocyclone signature can be used to give greater weight to the likelihood of a tornado. If an ETVS is detected, it can be considered to be rotation aloft that may develop into a tornado.

Some problems with using the TDA are that a volumetric scan is necessary to produce a TDA detection, so up to 6 minutes may pass before the forecaster learns of it. It is also very sensitive to problems inherent in the data, such as non-meteorological targets, ground clutter, beam broadening, and overshooting, and this may result in false alarms or no alarm at all. Finally it does not detect anticyclonic tornadoes, such as those that occur in a left split. In that respect, this algorithm, like all others, is a tool for forecasters to use in their own diagnosis of the situation, and is not an alerting system.

6.8. Hail detection algorithms

The WSR-88D uses an hail analysis algorithm known simply as Hail Detection Algorithm (HDA). It has gradually improved since the 1990s and now ingests the height of the 0°C and -20°C levels either from the radar operator or from the RR numerical model. The algorithm searches for high reflectivity values above the 0°C level, with 45 dBZ being a threshold value. The higher this is found above the 0°C level, the higher the probability of hail

Figure 6-5. Storm total precipitation for Hurricane Sandy as of early morning on 30 October 2012, as detected by the Mt. Holly, NJ radar. and viewed in GRLevel3 This shows a concentrated "hot spot" of over 10 inches of rain in the southernmost tip of New Jersey. It should be remembered that storm total precipitation is for a single site only, and may be distorted significantly by particles not comprised of rain. Future upgrades to NEXRAD will introduce better sensitivity to hydrometeor types for precipitation estimates.

(POH). There is also a probability of significant hail (POSH) which is based on how high the 40 dBZ reflectivities are relative to the 0°C and -20°C levels and whether higher reflectivities exist. Another quantity is maximum estimated hail size (MESH) which estimates the largest hail size in the cell.

The standard symbol used for HDA algorithm centroids is an *upward* pointing triangle. The NWS AWIPS system uses a large triangle to mark areas with a high POSH value. With GRLevelX and AWIPS, the number in the center corresponds to hail size in inches, and on AWIPS an asterisk is used for hail less than 3/4 inch in diameter. On AWIPS, a hollow triangle is used for hail probability that is less than a user-defined threshold, typically 50% by default.

This type of hail detection algorithm is degraded when storms are at a great distance from the radar, where the reflectivity has poor vertical resolution due to beam broadening, and close to the radar where high reflectivities in the upper parts of the storm lie within the cone of silence. As a result, the hail detection algorithm works best at a distance of 30 to 60 nm from the WSR-88D and less reliably outside this range. However, the hail algorithms have improved greatly since the 1990s and have matured into dependable components of the radar's forecast tools.

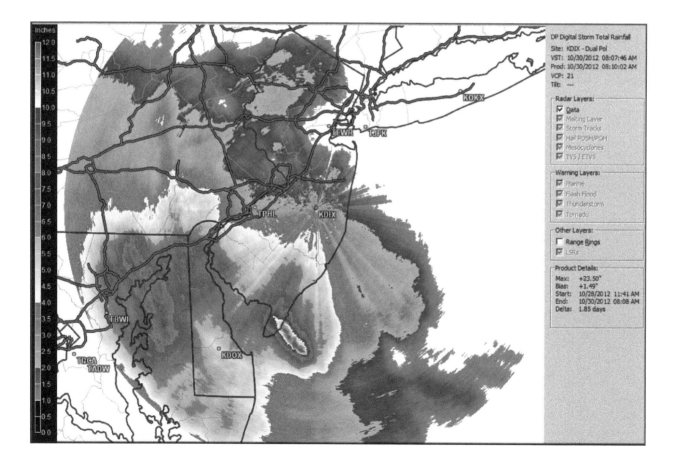

6.9. Storm relative velocity

The storm-relative velocity (SRV) product is derived from the base products and environmental winds. It is described in more detail in Chapter 4, velocity.

6.10. Velocity azimuth display

The velocity azimuth display (VAD) product is a graphical plot of radial velocity (x-axis) and azimuth (y-axis) at a given tilt. Velocity wind profiles (VWP) are sometimes erroneously described as this product on Internet sites and radar tutorials. Though the VAD product is disseminated continously at all radar sites, the actual images are rarely seen or used, and it serves as a source of data for the VWP product. See Chapter 4, velocity, for more information.

6.11. VAD wind profile

The VAD wind profile (VWP) is derived from the velocity azimuth display, which is also a derived product. It is described in more detail in Chapter 4, velocity.

REVIEW QUESTIONS

1. Describe how a composite reflectivity value for a given pixel is obtained.

2. Which will have a higher VIL value: intense precipitation through a shallow layer or intense precipitation through a deep layer?

3. What is the difference between base reflectivity and composite reflectivity?

4. Aside from detecting mesocyclones and tornadoes, what is the primary difference between the mesocyclone detection algorithm (MDA) and tornado detection algorithm (TDA)?

5. What is an advantage the mesocyclone detection algorithm has over the diagnosis of a mesocyclone based on a 0.5° storm-relative velocity image?

6. By convention what does a hollow TDA triangle indicate?

7. What type of shear does the TDA use as a basis for the identification of tornadic rotation?

8. Expand the acronyms POSH and MESH. Which one is a better discriminator for issuing a severe thunderstorm warning?

9. Why do the precipitation algorithms tend to work less reliably in sheared environments?

10. What time period does the WSR-88D storm total precipitation product cover?

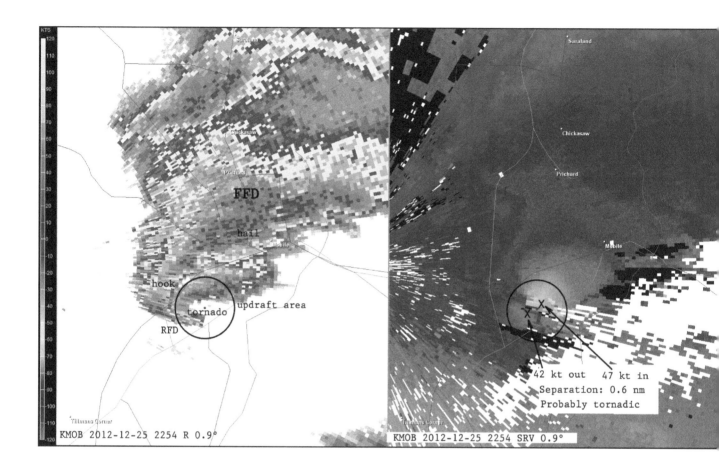

Figure 6-6. Mobile, Alabama tornado of Christmas Day 2012. The 0.9° reflectivity is shown at left and 0.9° storm relative velocity is shown at right. Because of the very close range of the radar, 0.9° was selected instead of 0.5° to eliminate a minor amount of ground clutter. Note how the maximum and minimum in the couplet was measured and their distance found from one another to determine what type of circulation it might be.

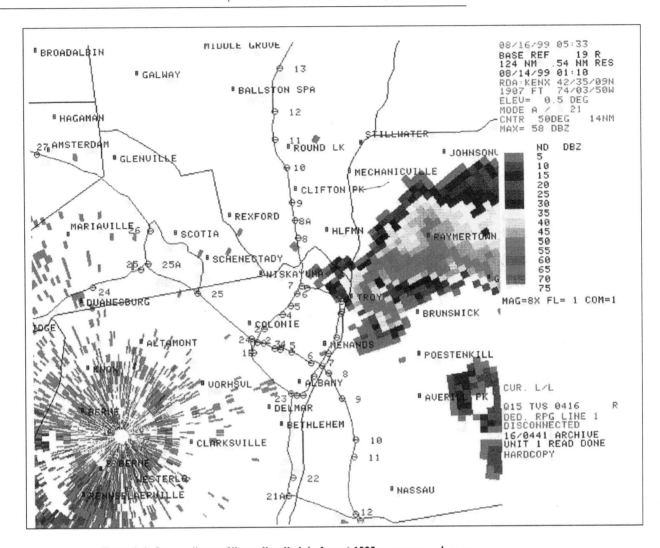

Figure 6-6. Supercell near Albany, New York in August 1999 as seen on a legacy WSR-88D printout. This style of graphic is often encountered in journals, Powerpoint slides, and online websites. It was produced by a Tektronix thermal wax printer at the WSR-88D PUP position. This type of printer would melt crayon-like blocks of wax and use it to print on paper and transparent slides. It is essentially a screenshot of one of the two PUP displays with a print request timestamp overlaid at the lower right.

7 Forecast Integration

In previous chapters, we started with equipment and fundamentals and considered radar-observed patterns and what they mean in terms of the underlying meteorological phenomena. Here we take the reverse, starting with meteorological patterns and identifying the correct use of radar products, along with concepts that a forecaster will find most useful. It must be emphasized that radar meteorology is at the pinnacle of a profoundly deep body of meteorological knowledge, and there is no way to build up even a working understanding of storm-scale analysis, hurricanes, and winter storms without adding hundreds of pages to this book. The reader is invited to consult some of the references cited in the appendix should any of the material become unclear. However the aspects of meteorology that are *directly* related to radar observation will be discussed.

Naturally, the first consideration in a meteorological situation is the big picture: the dominant type of weather situation and the overall long wave pattern as shown on 200 or 300 mb charts. Embedded within this is the synoptic scale and mesoscale weather regime, which is diagnosed with surface and low to mid tropospheric charts. This identifies boundaries and precipitation that will be found in the radar data. Since we have essentially produced a mental model of the atmosphere, all of the meteorological radar images have context.

7.1. Quiescent weather

Even when no active weather situations are taking place, a wealth of information can be gained from radars. Fronts and boundaries can easily be tracked using reflectivity products from the lowest tilts. They are also detectable to a certain extent on spectrum width products.

On days where storms are possible, a valuable task is to review regional radar mosaics over the past 24 hours. Going through 24 hours of data may seem as somewhat of a strange thing to do for forecasters who are accustomed to starting with current products and moving forward, but this task establishes where major convective weather systems have occurred, where mesoscale outflow pools are likely to have developed, and where outflow boundaries are likely to be found if they don't already appear.

In pre-storm environments, the wind profiles, especially in the lowest couple of kilometers, are extremely useful for the construction of manual hodographs. One reason for this is that 1200 UTC radiosonde wind data is usually obsolete by the time afternoon arrives. Even if a profiler station is already nearby, the radar wind profile can provide another data point with which to compare and contrast the profiler data. The VWP (Velocity Wind Profile) product is most useful for getting wind data from the WSR-88D, but this can be enhanced further by careful analysis of the zero line on base velocity..

Title image
Hurricane Katrina enters eastern Louisiana on 29 August 2005, at 1354 UTC. The radial velocity product showed winds of 63 kt (72 mph) near ground level. Higher tilts showed maximum radial winds of 125 kt (145 mph) at an altitude of 2500 ft MSL in the bands east of the radar site.

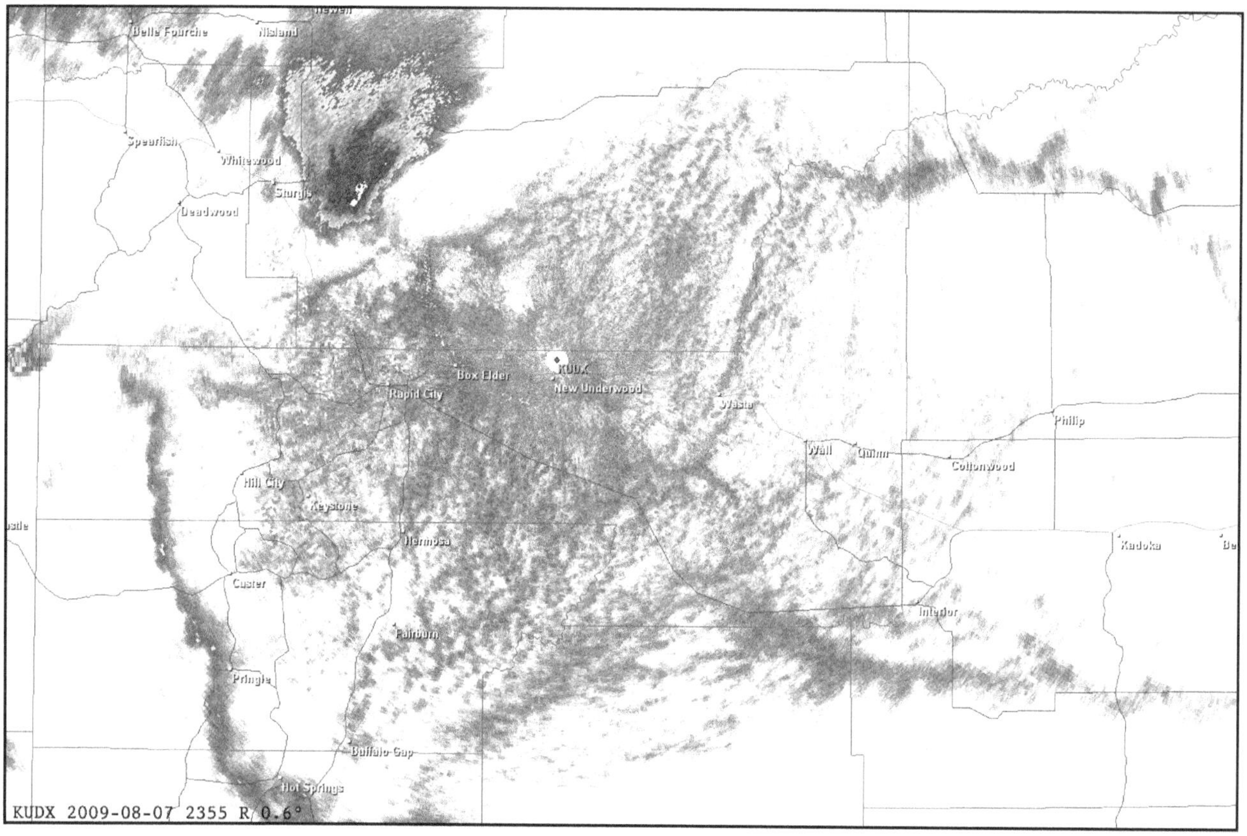

Figure 7-1. Outflow boundaries near Sturgis and Rapid City, South Dakota on 7 August 2009. All of these boundaries are favored areas for initiation or propagation of cells, including the supercell shown at top left. It had moved through Sturgis, dropping baseball sized hail. Supercells which propagate along well-developed boundaries have a greater chance at tornadogenesis compared to those that don't.

7.2. Boundaries

In radar meteorology, any zone of density or wind contrast demarcates an axis known as a boundary. Boundaries are often visible on radar as a "fine line", caused by convergence of particulates, water drops from convection along the front, and enhanced refractivity gradients.

7.2.1. FRONTS AND DRYLINES. Fronts form some of the strongest fine lines seen on radar data. To a lesser extent, drylines can also form fine lines, but when they are weaker or receding they may show no signatures at all. In terms of thunderstorm activity, fronts and drylines are important primarily in terms of serving as a source of convergence, which augments convective initiation and supports storm inflow. Storms that form along fronts which have significant temperature drops over a short distance will tend to have outflow dominant modes. This in turn may lead to either weak convective activity or HP supercells and the aggregation of them into a squall line.

7.2.2. OUTFLOW BOUNDARIES. These are critical features in mesoscale analysis before the development of convection. They may be a focus for convective initiation and may serve as a source of horizontal vorticity which heightens the risk of tornado development in any storm that later moves along the boundary. Therefore the forecaster has to examine the latest visible satellite imagery to assess whether storms are likely to form along the outflow boundary, and whether the boundaries are aligned in a way where right-movement vectors would favor propagation along the boundary.

If boundary axes are perpendicular to storm movement, then the outflow pool supporting the outflow boundary should be assessed to find where the cool outflow air is deepest and coolest, and determine whether this may cause weakening of the cell.

Stratiform VCPs
During stratiform precipitation events, the unit should be operating in VCP 21 or 121, which avoids most of the higher tilts and instead focuses on low tilts.

7.3. Stratiform precipitation

Stratiform precipitation is by its very nature fairly uniform throughout large scales of space and time. The areas show light to moderate in intensity and are of little vertical extent, rarely reaching more than 15 to 20 thousand feet in depth. Stratiform precipitation is comprised primarily of nimbostratus, altostratus, and cirrus cloud types and is typically produced by isentropic ascent, such as that over a warm front, or synoptic-scale destabilization of a humid layer.

If convective or slantwise instability increases beyond very low values, the stratiform precipitation may develop convective elements, often referred to as "embedded convection". These cells are made up of cumulonimbus clouds and are treated in accordance with standard thunderstorm interpretation techniques. Further destabilization or cell organization will cause the stratiform precipitation layer to transition into a multicellular convective system.

In a warm weather regime, interpretation of stratiform echoes is fairly straightforward. Winter weather, however, is a special situation and is described in detail in a later section of this chapter.

7.4. Convective precipitation

Radar is essential for a thorough analysis of severe convective storms, but on the same token severe convective storms are one the most important meteorological phenomena in radar meteorology. Because they go hand-in-hand, they were the catalyst for the United States upgrade to the WSR-88D radar. The fundamentals of thunderstorms as detected by radar were established in 1949 by the Thunderstorm Project. Though storm structure itself is outside the scope of this book, a summary of the basic concepts of storm classification is given here.

7.4.1. THE PRE-STORM ENVIRONMENT. Though the radar is known for its ability to detect severe weather, it has exceptionally useful capabilities for finding fronts and outflow boundaries, all of which signify a zone or axis of strong wind convergence that may favor development of a storm. Such boundaries may be detected and tracked many hours before storms actually form. Even if a fine line marking a boundary is not found, the location may be suggested by differences in the alignment of cloud streets detected by the radar.

7.4.2. THE DOWNDRAFT AND UPDRAFT. A storm consists of an updraft and a downdraft. The an updraft, which is manifested by the cumulonimbus tower, a downdraft, consisting of precipitation.

Since weather radars detect water drops, not cloud droplets, the initial cumulus cloud is invisible to radar until the cumulus cloud grows into towering cumulus. Here, droplets begin growing near the top of the cloud, typically at around 15,000 ft AGL (5 km) and become visible on radar. If the towering cumulus cloud grows into a cumulonimbus cloud, the radar echoes in the middle troposphere will expand and increase in intensity, and within 10 or 20 minutes will descend to the ground where they produce precipitation.

7.4.3. STORM TYPES. Before the 1950s, storms were grouped into three types: frontal, air mass, and squall lines. A better understanding of storm structure during the next few decades emerged, and storm types are now better described in terms of their updraft and downdraft structure. Storms with a single pair of updrafts and downdrafts are somewhat rare and are often associated with textbook models of a storm's life cycle, but a particularly noteworthy type of unicell storm is the supercell, a severe storm containing a highly persistent updraft-downdraft pair and unique structure. Storms with multiple pairs of updrafts and downdrafts are known as multicell storms, and they may range from disorganized complexes known as multicell clusters to organized systems like multicell lines and squall lines.

Once the storm enters a severe weather mode, it generally takes on three distinct features: very strong intensities, strong reflectivity gradients, and displaced cores. The strongest intensities will be in excess of 50 to 60 dBZ in a severe storm, much of this due to very intense rain and/or hail. The reflectivity gradients and displaced cores occur due to strong updrafts favoring one particular quadrant of a storm, and is usually associated with a tendency to propagate. The strongest severe storms will form distinct shapes, such as the triangular shape of a classic supercell.

7.4.4. HAIL. Perhaps the best indicators of hail come from the hail detection algorithm (HDA) product. In addition to this there are other indicators of hail. Very strong reflectivity is the primary indicator of hail, particularly when it is found at high altitudes (above 10 to 20

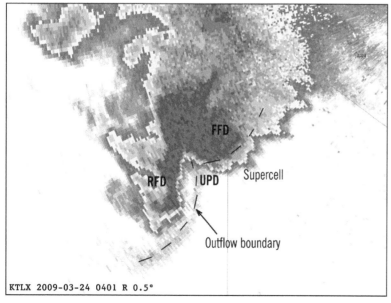

Figure 7-2. Near-storm outflow can be assessed to determine whether a storm is outflow dominant or not. In the case of the storm seen below, the outflow boundary is located underneath the updraft area, indicating the storm is inflow dominant or balanced, and still has the potential to produce a wide range of severe weather. If the storm outflow is "burped" out ahead of the storm, then the storm tends to become more outflow dominant and produces marginal severe weather or none at all. Outflow dominant storm outflow normally lies 5 to 15 miles ahead of the cell, and sometimes even further.

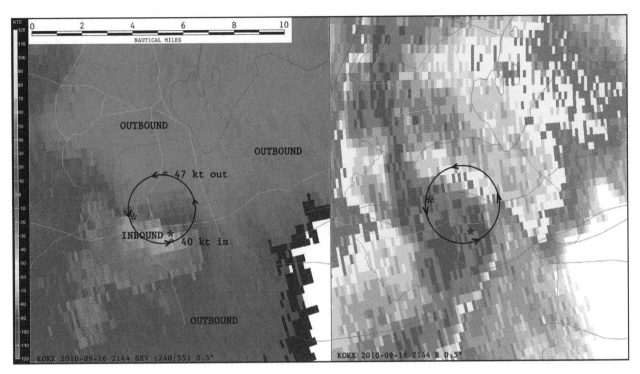

thousand feet). Polarimetric radar can also provide useful clues to the presence of hailstones. The three-body scatter spike is an artifact seen on reflectivity imagery and is associated with large wet hail.

7.4.5. HIGH WIND. Ground winds are monitored using the lowest available base velocity product. Bowing or rapid spreading of a cell on reflectivity may be one of the first signs of high wind. Though not always present, a localized rear inflow notch on the back side of a squall line may also be indicative of a potential area of high winds that may in turn progress to bowing of a cell and high winds at the ground.

7.4.6. MESOCYCLONES AND TORNADOES. The key product for locating and tracking storm circulations is the storm relative velocity product. The lowest elevation is normally monitored, but higher tilts should be checked regularly as circulations may form at higher levels or be detected better at higher tilts. The mesocyclone detection algorithm (MDA) and tornado detection algorithm (TDA) are highly useful for confirming the likelihood of a mesocyclone or tornado, but these should be used in conjunction with the velocity products. The reflectivity products are used for identifying the general structure of the storm and identifying likely areas for mesocyclone and tornado development. Signatures like hooks and appendages may occur, particularly on the rear quadrant of the storm facing the inflow (normally right rear in the northern hemisphere), but with high precipitation supercells the updraft area will show

Figure 7-3. Queens, New York tornado of 16 September 2010. Though it was rain-shrouded, its footage on YouTube attracted millions of views. The storm-relative velocity (left) and reflectivity (right) for 2144 UTC shows the distinctive signature of an HP supercell, with somewhat of a kidney bean shape on the reflectivity product and the mesocyclone (solid circle) being well embedded. This mesocyclone had a diameter of about 3 nm with about 85 kt of shear across this volume. The smaller-scale signatures of tornadoes are not easily found, though two locations of strong, localized shear are marked by asterisks within the mesocyclone. Part of the problem is that the system was 44 nm from the radar, and this was the closet network radar available.

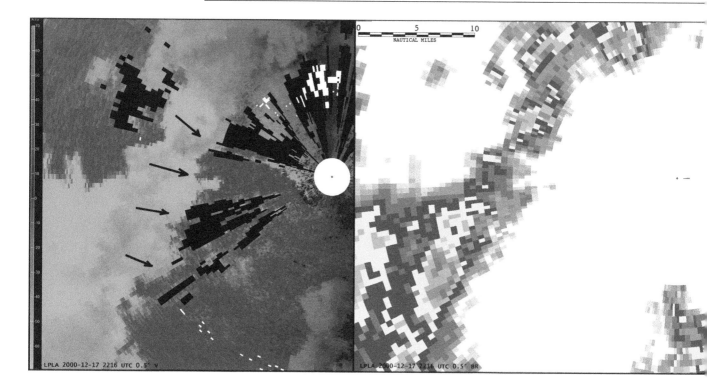

Figure 7-4. Windstorm event as detected by the WSR-88D at Lajes Field, Azores (Portugal) on 17 December 2000. The 0.5° radial velocity is shown on the left and 0.5° reflectivity on the right, indicating showers and weak thunderstorms. Placing the mouse on the leading edge of the area of strong inbound velocities showed widespread values of 45 to 50 kt (52 to 58 mph), contrasting sharply with the 10 kt of inbound flow ahead of it represented by the dark colors. The worst of the wind had gusted out by the time it reached Lajes Field, but gusts of up to 38 kt were recorded.

more of a notch or concavity in the forward or inflow quadrant of precipitation rather than a hook.

7.5. Winter weather

Winter weather situations pose an extra amount of complexity, as there can be changes in precipitation type both in the horizontal and in the vertical. The forecaster must have an understanding of the thermal structure of the atmosphere in the vertical, using sources of data like radiosonde and surface observations, and understand the different types of precipitation. The atmosphere is considered "cold" if it is below freezing, and "warm" if it is above freezing. From there, we can divide winter weather into several regimes based on the vertical temperature profile.

7.5.1. WARM ATMOSPHERE. If the atmosphere is entirely warm, only liquid precipitation will be possible. Water vapor condenses directly into liquid droplets, and these grow by collision and coalescence. Rain will fall at the surface.

7.5.2. COLD ATMOSPHERE. If the atmosphere is entirely cold, the production of ice crystals is favored, and snow is likely. Ice crystals develop by heterogeneous nucleation, and grow by deposition. The ice crystals may grow into snowflakes

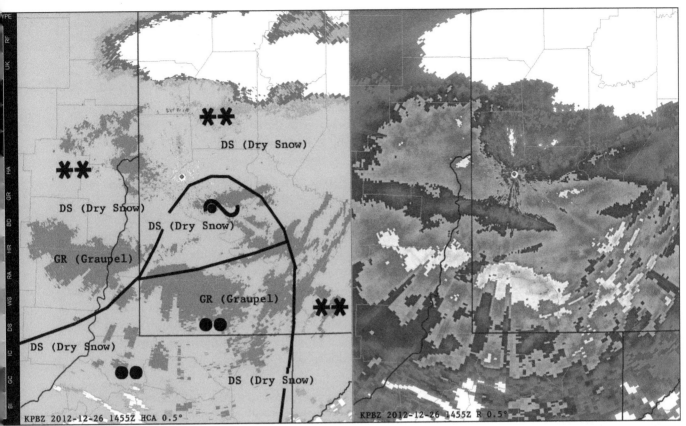

Figure 7-5. Example of an HCA failure during the 26 December 2012 winter storm in southwest Pennsylvania. The WSR-88D HCA algorithm showed mostly dry snow and graupel, even close to the radar site where the beam was near the surface and freezing rain was occurring while the HCA product showed dry snow. Actual surface reports (indicated by the weather symbols and bounded by thick solid lines) showed there were actually extensive areas of rain and freezing rain. Until HCA is refined and perfected, it should only be used as one of many tools. The forecaster must use all available data, including radiosonde, surface observations, and profiler data.

through aggregation and riming. These processes are described in detail in the margin.

However, it must be remembered that supercooled droplets can exist at temperatures well below freezing. In fact, if cloud temperatures are warmer than about -5°C, supercooled drops are likely to exist in the cloud rather than ice or snow, as contact with foreign particles fails to freeze the drops. It is only at temperatures of less than -5 to -10°C that ice is probable, and only at -15 to -20°C that it is a certainty.

The temperature of -15°C is exceptionally favorable for ice crystal growth due to the production of dendrites. These are branched types of ice crystals that easily aggregate into snowflakes. Furthermore, due to the differing fall speeds between heavy snowflakes and light ice crystals, collision and aggregation of ice crystals is enhanced. Because of this, the region between -12 and -18°C is known as the dendritic growth zone, and in the presence of significant lift, it will produce strong snowfalls.

7.5.3. Cold Above / Warm Below.

Much like the cold-atmosphere situation, snow will be the most likely precipitation type to enter the warm layer. Whether snow reaches the ground or not depends on the depth and strength of the warm layer. If the layer is slightly warm, precipitation will reach the ground as wet

Heterogeneous nucleation occurs at temperatures of about -10° to -20°C when water vapor condenses as supercooled water droplets. These are extremely small. If they contact another particle like dust, clay, pollen, or an existing ice crystal, they "flash freeze" in a process known as nucleation.

Accretion is a growth process for ice crystals caused by contact with microscopic supercooled droplets.

Deposition occurs when water vapor condenses directly onto an ice crystal as a liquid, and this liquid immediately freezes. It is an exceptionally common process in snow events. At colder temperatures water vapor may change directly into a solid ice particle.

Aggregation is the merging and sticking of ice crystals together to form a snowflake. This process is strong at temperatures above -5°C, especially near 0°C.

snow even if surface temperatures are well above freezing. In deeper, warmer layers, the snow will tend to melt out, producing mixed rain-snow or rain only.

The height of the melting level and its changes over time provides considerable information on the warm layer depth. The melting level is detected in reflectivity products as a bright band, and with polarimetric radar it best detected with the cross-correlation coefficient. With both products this appears as a ring around the radar site, owing to the conical nature of a radar tilt with increasing height corresponding to increasing range.

7.5.4. WARM ABOVE / COLD BELOW. Even in the warmest environments, the freezing level is rarely any higher than 20,000 ft, so this section actually refers to any elevated warm layer that is bounded by both cold air below (a shallow polar air mass) and cold air above (the upper troposphere). A deep warm layer, such as that found in the tropics, will be many thousands of feet deep and will exclusively form rain, but depending on the vertical extent of the cloud, ice crystals, snow, and graupel may also originate in the cold layer above. Likewise, in winter weather situations, the warm layer may be measured in mere hundreds of feet. Lift may result in rain within the shallow warm layer, but the vast majority of the precipitation production will be in the cold layer above, and the warm layer will be filled with snow falling from above.

Before even considering the low-level cold layer, the forecaster must determine what type of precipitation will be occurring at the bottom of the elevated warm layer. If it is deep, only rain will exist since any snow will have melted. On the other hand, if it is extremely shallow, the warm layer may have a negligible effect on precipitation type and the situation can be treated as a "cold atmosphere" situation. Many winter weather events fall somewhere in between. Roughly 600 ft of warm layer depth is sufficient to melt snowfall, but an exact value depends on the temperature and humidity in the warm layer.

Using this estimate of precipitation type at the bottom of the warm layer, the forecaster then estimates the effect of the low-level cold layer on this precipitation. If the cold layer is sufficiently shallow, rain will freeze upon contact with objects on the ground, producing freezing rain. With deeper cold layers, the drops will freeze as they fall, resulting in ice pellets (sleet) at the surface. As a general rule of thumb, if rain is entering the cold layer, ice pellets will be the result if it has more than 800 ft of depth and freezing rain if it is shallower, but this is only a typical value and should not be relied upon.

The forecaster might also have concluded that snow, not rain, will be entering the top of the cold layer. This greatly simplifies the forecast since the snow will remain frozen in the low-level cold layer.

It must be remembered that warm-over-cold situations are where polarimetric products and the hydrometeor classification algorithm (HCA) are most likely to fail, since it does not take into account this low-level cold layer. The temperature characteristics of the lowest 500 to 1000 ft of the atmosphere are critical in shallow cold layer events, and since this normally lies well beneath

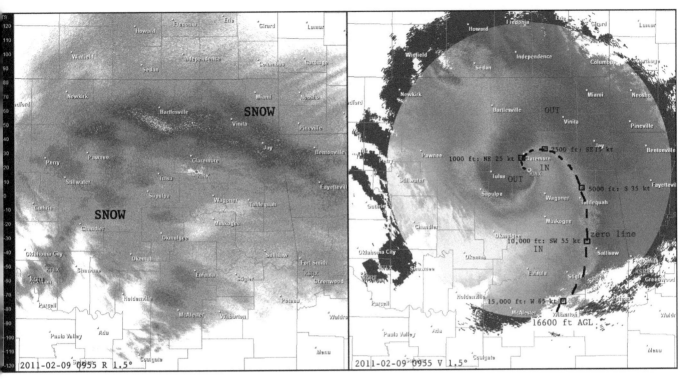

Figure 7-6. **Basic winter weather interpretation** using base reflectivity (left) and velocity (right). The soft textured patterns on reflectivity indicates snow, and this is confirmed by surface observations. The velocity product on the right yields insight into the depth of the cold air mass. By follow the zero line from the surface to the edge of the volume, we see that the winds start out northeasterly at the surfae, switching to southerly and then westerly. This can yield information about cold air mass depth and whether it is increasing or decreasing.

the lowest available tilt, the radar is normally not able to sample this critical layer. It is up to the forecaster to make a final determination of how the cold layer will affect precipitation type. Regardless of what the HCA product shows, cold rain, ice pellets, freezing rain, or snow may be experienced at the surface depending on the depth and intensity of the cold layer.

7.5.5. SNOWFALL. The amount of snowfall that occurs is simply the intensity multiplied by the time elapsed. Duration of precipitation is well-forecast by numerical models, but intensity tends to be poorly resolved, and here is where radar can provide additional information.

7.6. Tropical cyclones

The Doppler dilemma is major problem for third-generation radars sampling tropical cyclones. This is because the precipitation fields can be very extensive, covering the entire radar surveillance area at great ranges while containing very high velocities. No matter which PRF is selected, there is always a tendency for both range folding and velocity aliasing to occur.

For the WSR-88D, the special MPDA (multi-PRF dealiasing algorithm) which uses different PRFs to maximize both the maximum unambiguous range and maximum unambiguous velocity is paramount. This is implemented in VCP 121, so this VCP is widely used when tropical cyclones make landfall. The forecaster may instead elect to sacrifice dealiasing ability in order to eliminate

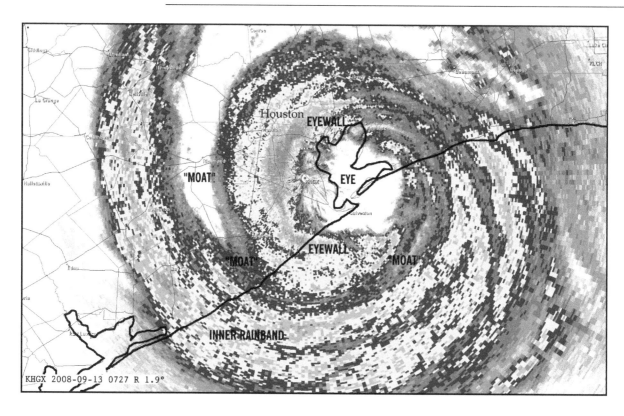

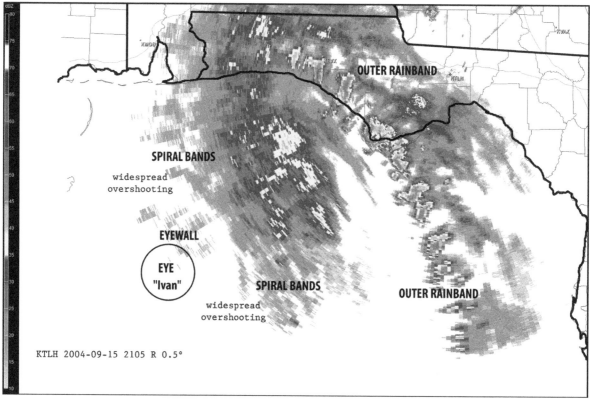

the purple ring of range folded data surrounding the radar site about 65 nm away. This is done using VCPs 211, 212, and 221, which have the special SZ-2 range unfolding algorithm. Out of these three, VCP 221 is the best for distant hurricanes since it has very few high tilts. When hurricanes are closer, VCP 211 is preferred since this introduces higher tilts.

The strength of the hurricane can be measured by simply querying the velocities shown at 0.5° and looking for the strongest consistent values of base velocity. Since this only measures the radial component of the wind, the actual winds are higher than the values shown unless the radial is looking into the flow head-on.

Tornadoes are responsible for about 10 percent of the deaths in hurricanes. Most of these tornadoes occur in the outer rainband, which usually precedes the hurricane by 100 to 200 nm in the northeast quarter of the storm (southeast in the southern hemisphere). These tornadoes occur within very shallow depths and are rather transient, so it is normally difficult to identify and track them. Storm relative velocity has been proven to be a very useful product for monitoring a tornado's velocity couplets. The study by Spratt et al (1997) suggested that attempting to distinguish thresholds of rotational velocity in these couplets is not worthwhile and forecasters may have better luck simply tracking the couplets. Spectrum width may be useful in helping to do this.

Figure 7-7a (facing page, above). Hurricane Ike on 13 September 2008 as it moves ashore near Houston. Here the main parts of the tropical cyclone located near the circulation are identified. The outer rainband is too far away to appear here.

Figure 7-7b (facing page, below). Hurricane Ivan makes landfall on 15 September 2004. It was the most prolific tornado-producing hurricane to strike the United States., producing 127 tornadoes, 18 of them being F2 or stronger, though most of these were in the Appalachian region and in the Northeast. This image shows the outer rainband, which is a favored location for tornadogenesis. This particular spiral band produced 26 tornadoes over the following six hours. Note that the inner rainbands and the eyewall itself are out of range.

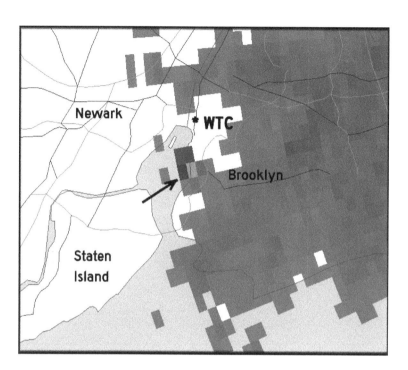

Figure 7-8. On 11 September 2001 we see New York City about 13 minutes after the collapse of the South Tower of the World Trade Center. This 32 dBZ echo was the first strong signature to be detected, but with the radar being 53 nm away it took over ten minutes for the smoke to be lofted high enough where it could be detected by the 0.5° tilt (in this case 4800 ft MSL). This plume also had drifted 3 miles downwind. In this respect the radar can be useful for monitoring smoke and hazardous chemical trajectories, if scatterers are significant enough to be detected. It should be noted that base reflectivity has improved significantly since 2001 due to the introduction of super resolution data, and coarse images like this have become somewhat of a rarity.

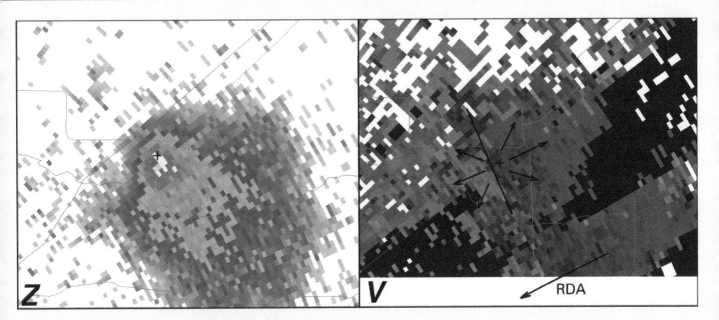

Figure 7-9. Dramatic example of biological backscattering as observed by the Little Rock, Arkansas WSR-88D on December 31, 2010 at 11:25 pm CST. New Years firework explosions in the southern part of Beebe, Arkansas (marked with a cross) caused thousands of blackbirds and starlings to fan out across nearly 100 square miles. Both frames are for the 0.5° tilt, so this crosses over Beebe at about 1000 ft AGL. The reflectivity product (left) showed a bloom of backscattered energy from the birds, and the radial velocity product (right) shows a distinct divergence signature (zero-line in solid black) caused by dispersion away from the disturbance, contrasting sharply with the light westerly flow across the region.

This pattern does have the resemblance of small convection producing outflow, and indeed showers did move through earlier in the day, but reflectivity frames showed a series of dispersions originating from a specific location (within about a mile of 35.06,-91.90) at 10:17 pm, 10:55 pm, 11:16 pm and 11:55 pm. Repeated appearances of significant reflectivity at a fixed location are often associated with wildfires. However since wildfire particles tend to drift with the existing wind fields rather than strongly diverge, velocity products rarely show any identifiable signature. With larger wildfires, pyrocumulus develops and may grow into cumulonimbus, producing the usual expected range of meteorological signatures associated with precipitation. In this case, neither a wildfire nor any precipitation was occurring.

The fact that no biological scatterers were detected elsewhere around the state, even at midnight, and that fireworks were likely commonplace across the region is suggestive of an exceptionally high bird population, either due to migration or overpopulation, the presence of exceptionally powerful, numerous, or deliberately targeted fireworks in Beebe, or even a combination of the two.

The 2010 New Years Eve incident was quick to appear in state news headlines. Television journalists reported the next day that workers were cleaning up thousands of dead birds. The Arkansas Game and Fish Commission reported that the birds were most likely spooked from their roosts then injured while in flight. There has been general consensus among ornithologists that the birds were disoriented in the dark and hit trees, structures, each other, or even the ground itself. Not surprisingly, the continuous media references to "birds falling out of the sky" have also raised an equal share of conspiracy theories ranging from lightning to toxic gas clouds.

In a surprising turn of events, the same New Years Eve phenomenon was observed again on the evening of December 31, 2011. The Little Rock WSR-88D, which was operating in clear air mode, detected a bloom of echoes at the same location as last year beginning at 7:06 pm. An Associated Press report said that when the first reports came in of dead birds at 7 pm the town initiated an emergency ban on fireworks. However, this apparently had limited effect. A larger bloom appeared midnight over much of the town, though it was smaller than the previous year's signature, and it was reported that only about 200 dead birds were collected rather than the thousands of the previous year.

The city and the Arkansas Game and Fish Commission agreed that malicious fireworks use was the cause of the events and they had already been traced to a roosting area near Beebe's Windwood subdivision, corresponding precisely to the information obtained from the radar output. After the previous year's events police department had officers positioned in the subdivision ready for the possibility of problems, but were unable to locate those who were responsible.

gone with the wind
A study of the final images from destroyed WSR-88D RDAs

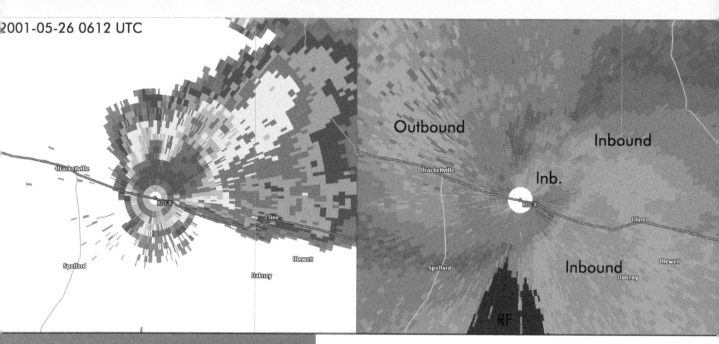

2001-05-26 0612 UTC

May 2001: Del Rio, Texas

Shortly after 1 a.m. CDT on the morning of 26 May 2001 a northwesterly flow supercell damaged the Del Rio, Texas (KDFX) RDA. The very last image (top left) showed an unusual "donut" artifact on the reflectivity image engulfing the RDA site. This donut may be due to three-body scattering by hail just outside the radome. The velocity image at the top right tells a more meaningful story, showing a zone of 55 knots (63 mph) area of inbound velocity immediately to the northeast, which probably marks the arrival of a downburst. If this sector is contaminated by three-body scattering then the actual velocities in those locations may be even higher.

The Del Rio antenna and radome had to be completely replaced. The radar was finally put back online 19 July 2001, marking the end of a 54-day outage.

```
NOUS64 KEWX 260612
FTMDFX
THE RADAR WILL BE DOWN UNTIL FURTHER NOTICE
DUE TO AN UNEXPECTED OUTAGE. SORRY FOR THE
INCONVENIENCE. 26/7Z
```

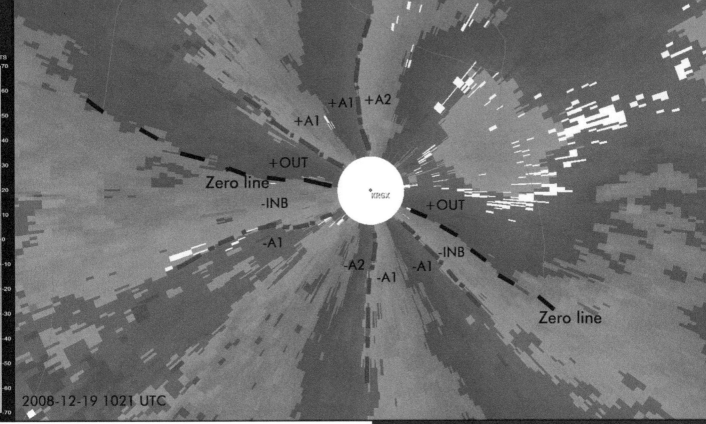

December 2008: Reno, Nevada

The Reno, Nevada (KRGX) radome was destroyed on 19 December 2008 during a severe Pacific winter storm. The last base velocity image (above) shows a bizarre pinwheel shape, but actually this is the effect of multiple velocity aliasing due to low ambiguous velocities and high wind speeds. The radar was operating in VCP 31, with a maximum unambiguous velocity of 22 kt. The METAR reports from Reno showed winds from 180° at 23 gusting to 34 kt, which implies an east-west zero line. Starting from this zero-line, careful analysis of velocities at a selected range will show "flipflops" in the sign of strong velocity at the dash-dot lines, revealing aliased sectors (labelled with the letter A). The double aliasing indicates that winds were well in excess of 44 kt at nearly all low-level and mid-level altitudes. It should be noted that GRLevel2's dealiasing algorithm was not able to "unfold" this particular volume and the result had the above pinwheel appearance, underscoring that forecasters must know how to recognize when de-aliasing algorithms fail. The radar was repaired and re-commissioned on 7 February 2009, ending a 50-day outage.

Appendix

Common radar abbreviations and variables

BR	Base reflectivity
BV	Base velocity
D	Diameter of precipitation particles
D_0	Median diameter of precipitation particles
D_m	Mass-weighted diameter of precipitation particles
ET	Echo top height
dBZ	Decibels of effective reflectivity (Z_e); the common expression of intensity or reflectivity
K_{DP}	Specific differential phase: $d\phi_{dp} / dr$L, usually given in deg km^{-1}
LDR	Linear depolarization ratio: $10 \log (Z_{HV}/Z_{HH})$
PHI	Differential phase (ϕ_{dp})
POSH	Probability of significant hail
R	Rainfall rate
r	Range (distance between antenna and target)
RHO	Cross-correlation coefficient of horizontal and vertical waves (ρ_{HV})
SRV	Storm relative velocity
SW	Spectrum width
W	Liquid water content
VIL	Vertically integrated liquid
Z_{HH}	Power, horizontal phase transmitted, horizontal phase received (co-polar); this is most common
Z_{VV}	Power, vertical phase transmitted, vertical phase received (co-polar)
Z_{HV}	Power, horizontal phase transmitted, vertical phase received (cross-polar)
Z_{VH}	Power, vertical phase transmitted, horizontal phase received (cross-polar)
V_r	Radial velocity
W	Spectrum width: $(\sigma^2)^{1/2}$
ZDR	Differential reflectivity
Z	Uncorrected reflectivity (an expression of electromagnetic power)
Z_e	Effective reflectivity
Z_{DR}	Differential reflectivity: $10 \log (Z_{HH}/Z_{VV})$
λ	Wavelength (lambda)
ρ_w	Rain water density
ρ_{HV}	Cross-correlation coefficient of horizontal and vertical waves: $\rho_{HV}(0) \exp^{j\delta}$ (rho)
σ	Spectrum width (sigma)
ϕ_{dp}	Differential propagation phase: $\phi_H - \phi_V$ (phi)

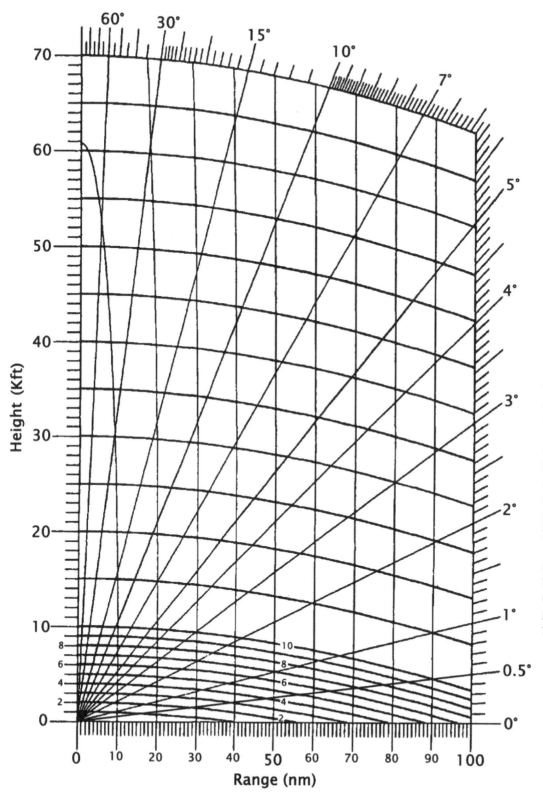

Standard radar range-height diagram for general reference use. It assumes a standard atmosphere. Most computer-based radar viewers such as AWIPS and GRLevelX will work this out interactively, displaying height anywhere the mouse or cursor is positioned. But for basic Internet images, mobile device displays, and just for visualizing the basics of a radar beam, a diagram like this is very helpful.

To use it, start from the range (distance) of the target of interest, move up vertically until the correct tilt (elevation) is intersected, then read the height using the curved horizontal lines.

This diagram is also useful for visualizing the width of the beam. The WSR-88D's half power points are about 1° apart, so at 0.5° the beam will sample a swath roughly between 0 and 1.0°. Looking at this 0 to 1° swath on the diagram, we see that at 100 nm the beam is over 10,000 ft (2 miles) wide.

All heights on this diagram are relative to the radar antenna, so essentially this diagram shows height above ground level (AGL) relative to the radar site.

THIS PAGE MAY BE REPRODUCED WITHOUT RESTRICTION
WEATHER RADAR HANDBOOK ©2013 WEATHER GRAPHICS TECHNOLOGIES

Storm motion nomogram. This diagram is useful for determing exact motion of storm cells, outflow boundaries, mesocyclones, couplets, and other features, using a couple of radar frames at different times. It is especially useful for setting specific storm motion values for user-defined storm relative velocity. Simply determine the time interval and measure the distance carefully, using a good radar display application. Starting on the left edge, proceed horizontally until the vertical time interval line is reached. The diagonal lines will yield the storm motion value. Any system of units may be used as the resulting number will be the distance the feature will move in one hour, so readers may use nm for distance and kt for velocity, or km for distance and km h-1 for velocity. Multiply km h-1 by 0.2777 if m s-1 is desired.

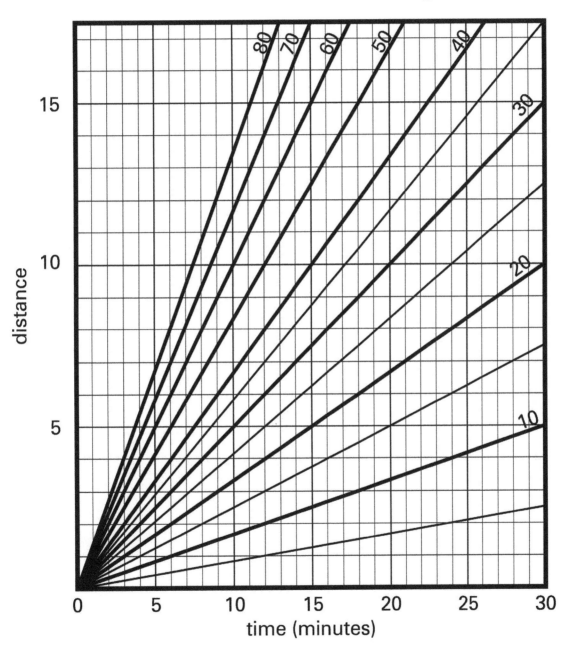

THIS PAGE MAY BE REPRODUCED WITHOUT RESTRICTION
WEATHER RADAR HANDBOOK ©2013 WEATHER GRAPHICS TECHNOLOGIES

NEXRAD product codes

NEXRAD Level III product codes

<u>FIRST-SECOND-THIRD LETTER COMBINATION</u>
DAA Digital accumulation array
DHR Digital hybrid scan reflectivity
DPR Instantaneous precipitation rate
DSP Digital storm total precipitation (HR)
DTA Digital storm total precipitation
DU3 Digital 3 hour accumulation
DU6 Digital 6 hour accumulation
DOD Digital one-hour difference
DSD Digital storm total difference
DVL Digital vert. integ. liquid (HR)
EET Enhanced echo tops
HHC Hybrid scan hydrometer class. alg.
NCR Composite reflectivity
NET Echo tops
NMD Mesocyclone
NST Storm tracking information
NVI Vertically integrated liquid
NVW VAD wind profile
NTP Storm total precipitation
NOx 0.5 deg (for x see THIRD LETTER table)
N1x 1.5 deg (for x see THIRD LETTER table)
N2x 2.4 deg (for X see THIRD LETTER table)
N3x 3.4 deg (for X see THIRD LETTER table)
NAx 0.9 deg (for X see THIRD LETTER table)
NBx 1.8 deg (for X see THIRD LETTER table)
OHA One hour precipitation
PTA Storm total accumulation
SPD Supplemental precipitation data
TZL Reflectivity 0.6 deg long range
TR0 Reflectivity base elevation
TR1 Reflectivity 1.0 deg elevation
TR2 Reflectivity third elevation
TV0 Velocity base elevation
TV1 Velocity 1.0 deg elevation
TV2 Velocity third elevation

<u>THIRD LETTER</u>
C Correlation coefficient (DP)
H Hydrometeor classification algorithm (DP)
K Specific differential phase (DP)
M Melting layer (DP)
Q 248 nm base reflectivity (HR)
R 124 nm base reflectivity
S Storm relative velocity
U 162 mile base velocity
V Base velocity
X Differential reflectivity (DP)
Z 248 nm base reflectivity

Summary of typical polarimetric parameters

Summary of typical polarimetric parameters of various types of water, ice, solids, and other scatterers. The K_{DP} values here are specific to the WSR-88D.

Type	Z (Z_{DR})	Z_{DR}	K_{DP}	CC (rho)	LDR
Drizzle	10-25	0.25 to 2	0 to 1	Low	Low
Light rain	20-35	2 to 4	1 to 2	Medium	Low
Heavy rain	35-50	4 to 7	3 to 10	High	Low
Ice crystals: needles	0-20	0.5 to 2	0-2	High	Low
Ice crystals: columns	0-25	2 to 3	0-2	High	Low
Ice crystals: plates	0-25	3 to 5	0-2	High	Low
Ice crystals: dendrites	0-25	-0.5 to 0.5	0-2	High	Low
Dry snow	15-3	-0.3 to 1.5	-1 to 0	Very high	Low
Wet snow	25-50	-0.5 to 3	-0 to 0.5	Low	Low
Graupel, dry	25-45	-0.5 to 1	0 to 1	High	Low
Graupel, wet	40-55	1 to 2	1 to 2	Low	Medium
Small hail, dry	40-50	-0.5 to 1	0 to 0.5	Medium	Medium
Small hail, wet	50-60	-2 to 1	-0.5 to 0.5	Low-medium	High
Large hail, dry	45-70	-0.5 to 1	-1 to 1	Low	High
Large hail, wet	60-80	-2 to 0	-1 to 1	Very low*	Very high
Insects and birds	10-20	1 to 8	-	Low	Low
Chaff	20-30	1 to 8	-	Very low	Very high
Debris	10-40	0.25 to 2	-	Very low	Very high
Ground clutter	-	-4 to 2	-	Very low	Very high
Clear air	-	-	-	Very low	-

Table 5-1. **Summary of typical polarimetric parameters** of various types of water, ice, solids, and other scatterers. The K_{DP} values here are specific to the WSR-88D.
* = due to Mie scattering
CCC VALUES: Very low=0-0.6, low=0.6-0.94, medium=0.95, high=0.96-1
LDR VALUES Low= <-30, medium=-29 to -25, high>-24
ICE CRYSTALS: Typical habitats for needles are -4 to -6°C; for columns -6 to -10°C; for plates -10 to -22°C; and for dendrites -12 to -16°C.

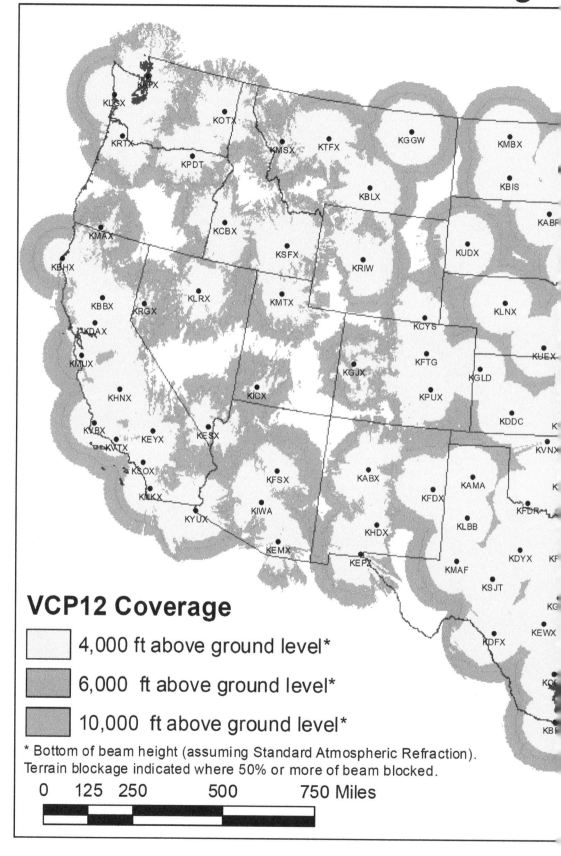

Below 10,000 Feet AGL

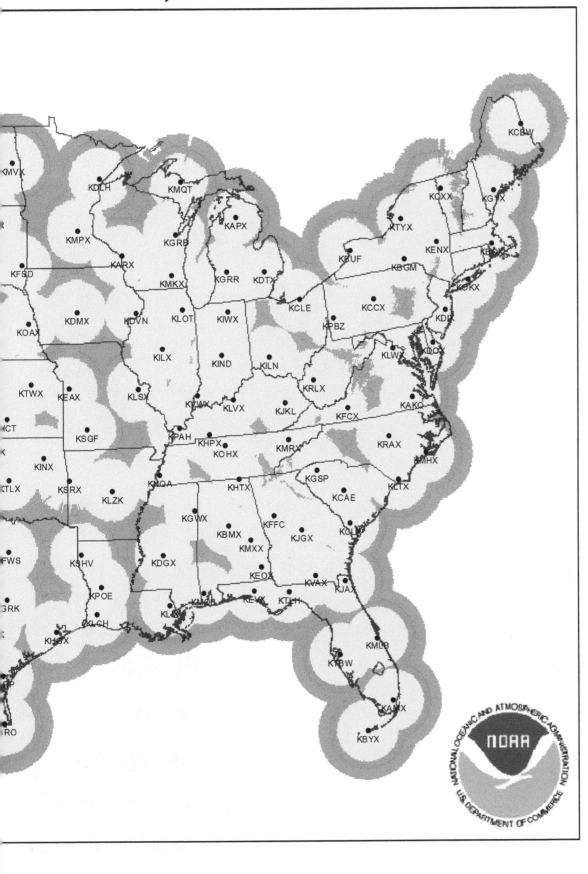

List of WSR-88D builds

The WSR-88D is a modular radar system that has been upgraded substantially over the years using hardware and software packages called "builds". This listing chronicles the build names, their date of operational release, and a short outline of the improvements. It is especially important to note that there was a changeover in 2002 with the implementation of Open RPG and Open RDA architecture, which restarted the build number sequence. This may be a source of confusion in older references.

Legacy Build 5 (May 1990)
Deployed to the OKC prototype radar.

Legacy Build 6 (May 1993)

Legacy Build 7 (1994)

Legacy Build 8 (March 1995)

Legacy Build 9 (September 1996)
Adds Storm Cell Identification and Tracking (SCIT), Hail Detection Algorithm (HDA), and CT.

Legacy Build 10 (October 1998)
Implemented a new Tornado Detection Algorithm (TDA), developed by NSSL. It differentiates between a TVS and elevated TVS, focuses on gate-to-gate velocity differences, does not require a mesocyclone, and offers adaptable parameters.

Open Build 1 (April 2002 - July 2002)
Based upon Legacy Build 10, this implemented the Open Build architecture. The most prominent enhancement for users is the upgrade of base products from 16 levels of intensity and velocity to 256.

Open Build 2 (October 2002)
Introduced super ob for NCEP, a rainfall bias table, and an REC clutter likelihood algorithm.

Open Build 3 (March 2003)
Added high-resolution VIL, digital storm total precipitation, user-selectable layer reflectivity, hodograph from VWP, and rainfall bias table.

Open Build 4 (September 2003)
Added enhanced echo tops, legacy meso rapid update, and precipitation rate/accumulation for new VCPs.

Open Build 5 (March 2004)
Added Mesocyclone Detection Algorithm (Phase 1), Tornado Rapid Update, new VCPs with 4.2 minute volume scans, Level II central collection, and AWIPS datastream upgrades.

Open Build 6 (September 2004)
Added Mesocyclone Detection Algorithm (Phase 2) and compression of some products.

Open Build 7 (June 2005)
Mostly data dissemination changes: increased Class 2 user access; increased WAN OTR flow control, and expanded RPS list size.

Open Build 8 (March 2006)
Added WAN ports and flow control, SCIT filter, Level III status product, and products added to default generation list.

Open Build 9 (June 2007)
Added SZ-2 ambiguity algorithm, Mesocyclone Detection Algorithm (Phase 3), environmental data ingest from the RUC, intelligent gust front algorithm.

Open Build 10 (May 2008)
Introduced **super-resolution radar** in the Level II datastream (0.5°x0.25 km) for base products.

Open Build 11 (May 2009)
Dual pol(?) Mesocyclone Detection Algorithm (Phase 4), environmental data ingest from the GFS.

Open Build 12 (September 2011)
Implements **dual polarization**, but uses contractor source code that reverts the other radar functionality to Build 10.

Open Build 13 (planned for 2013)
This is a government build that merges the contractor's dual polarization algorithms with government Build 11. This re-implements the Automated Volume Scan Evaluation and Termination (AVSET) and clutter mitigation decision (CMD) algorithm. It also introduces 2DVDA, the 2-Dimensional Velocity Dealiasing Algorithm, which improves dealiasing of data and reduces the number of false mesocyclone detections.

Build 14 (planned for 2014)
Will add an **extra 0.5° scan** during VCP 12 or 212 volume scans, providing 2-minute resolution at that tilt. Will add automatic PRF selection and a number of RDA improvements.

WSR-88D Volume Coverage Pattern (VCP) reference

WSR-88D Volume Coverage Pattern (VCP)

VCP	Mode	Tilts	Enh. range unfld	Volume scan time (min)	Pulse length (ms)	
11	A	14	-	5		
12	A	14	Y	4.5	Deep convection, faster than VCP11	
21	A	9	-	6	Shallow precipitation; fewer tilts	
121	A	9	Y	5.75	Offshore hurricanes, nonsevere weather	
31	B	5	-	10		
32	B	5	-	10		
211	A	14	Y	5	SZ-2(2)	
212	A	14	Y	4.5	SZ-2(3)	
221	A	9	Y	6	SZ-2(2)	

Mode: A is storm mode, B is clear air mode.

Enh range unfld: Enhanced range unfolding.

Pulse length: Pulse length in all modes is 1.57 ms (0.5 km, short pulse) except for VCP 31, which is 4.7 ms (1.4 km, long pulse).

Nyquist velocity: 8 to 32.8 m s^{-1} (16 to 64 kt) on all VCPs except VCP 31 (8 to 12.4 m s^{-1}, i.e. 16 to 24 kt) and VCP 32 (8 to 28.2 m s^{-1}, i.e. 16 to 55 kt)

Pulse repetition frequency (PRF): 318 to 1304 Hz on all VCPs except VCP 31 (318 to 452 Hz)

Remarks: SZ-2(2) is Sachidananda-Zrnic SZ-2 algorithm on the lowest 2 tilts, while SZ-2(3) is the same algorithm on the lowest 3 tilts.

Quick Reference VCP Comparison Table for RPG Operators

February 2007

Slices	Tilts	VCP	Time*	Usage	Limitations
19.5° 16.7° 14.0° 12.0° 10.0° 8.7° 7.5° 6.2° 5.3° 4.3° 3.4° 2.4° 1.5° 0.5° 0.0°	14	11	5 mins	Severe and non-severe convective events. Local 11 has Rmax=80nm. Remote 11 has Rmax=94nm.	Fewer low elevation angles make this VCP less effective for long-range detection of storm features when compared to VCPs 12 and 212.
		211	5 mins	Widespread precipitation events with embedded, severe convective activity (e.g. MCS, hurricane). Significantly reduces range-obscured V/SW data when compared to VCP 11.	All Bins clutter suppression is NOT recommended. PRFs are not editable for SZ-2 (Split Cut) tilts.
19.5° 15.6° 12.5° 10.0° 8.0° 6.4° 5.1° 4.0° 3.1° 2.4° 1.8° 1.3° 0.9° 0.5° 0.0°	14	12	4 ½ mins	Rapidly evolving, severe convective events. Extra low elevation angles increase low-level vertical resolution when compared to VCP 11.	High antenna rotation rates decrease the effectiveness of clutter filtering, increase the likelihood of bias, and slightly decrease accuracy of the base data estimates.
		212	4 ½ mins	Rapidly evolving, widespread severe convective events (e.g. squall line, MCS). Increased low-level vertical resolution compared to VCP 11. Significantly reduces range-obscured V/SW data when compared to VCP 12.	All Bins clutter suppression is NOT recommended. PRFs are not editable for SZ-2 (Split Cut) tilts. High antenna rotation rates decrease the effectiveness of clutter filtering, increase the likelihood of bias, and slightly decrease accuracy of the base data estimates.
19.5° 14.6° 9.9° 6.0° 4.3° 3.4° 2.4° 1.5° 0.5°	9	21	6 mins	Non-severe convective precipitation events. Local 21 has Rmax=80nm. Remote 21 has Rmax=94nm.	Gaps in coverage above 5°.
		121	6 mins	VCP of choice for hurricanes. Widespread stratiform precipitation events. Significantly reduces range-obscured V/SW data when compared to VCP 21.	PRFs are not editable for any tilt. Gaps in coverage above 5°.
		221	6 mins	Widespread precipitation events with embedded, possibly severe convective activity (e.g. MCS, hurricane). Further reduces range-obscured V/SW data when compared to VCP 121.	All Bins clutter suppression is NOT recommended. PRFs are not editable for SZ-2 (Split Cut) tilts. Gaps in coverage above 5°.
4.5° 3.5° 2.5° 1.5° 0.5° 0.0°	5	31	10 mins	Clear-air, snow, and light stratiform precipitation. Best sensitivity. Detailed boundary layer structure often evident.	Susceptible to velocity dealiasing failures. No coverage above 5°. Rapidly developing convective echoes aloft might be missed.
		32	10 mins	Clear-air, snow, and light stratiform precipitation.	No coverage above 5°. Rapidly developing convective echoes aloft might be missed.

*VCP update times are approximate.

Source: NOAA / National Weather Service / Warning Decision Training Branch

References & recommended reading

American Meteorological Society, 1990: Radar in Meteorology (Battan Memorial Volume). American Meteorological Society, Boston.

Battan, L., 1973: *Radar Observation of the Atmosphere.*

Bringi and Chandrasekar, 2001: *Polarimetric Doppler Weather Radar.* Cambridge University Press.

Brown, Rodger A., Vincent T. Wood, Dale Sirmans, 2000: Improved WSR-88D Scanning Strategies for Convective Storms. *Wea. Forecasting*, **15**, 208–220.

Caylor, I.J., and V. Chandrasekar, 1996: Time-varying ice crystal orientation in thunderstorms observed with multiparameter radar. *IEEE Trans. Geosci. Remote Sensing*, **34**, 847-858.

Crum, Timothy D. and Ron L. Alberty, 1993: The WSR-88D and the WSR-88D Operational Support Facility. *Bull. Amer. Met. Soc.*, **74**, 1669-1687. An excellent overview of NEXRAD development from the 1980s into the early 1990s.

Doviak, Richard J. and Dusan S. Zrnic, 1993: *Doppler Radar and Weather Observations.* Academic Press, 562 pp. A thorough, technical overview of the basics of Doppler radar.

Hoffman, R. R., M. Detweiler, J. A. Conway, and K. Lipton, 1993: Some Considerations in Using Color in Meteorological Displays. *Weather and Forecasting*, **8**, 505-518.

IIPS Subcommittee for Color Guidelines, 1993: Guidelines for Using Color to Depict Meteorological Information. *Bull. Amer. Metr. Soc.*, **74**, 1709-1713.

Krehbiel, P., T. Chen, S. McCrary, W. Rison, G.Gray, and M. Brook, 1996: The use of dual channel circular-polarization radar observations for remotely sensing storm electrification. *Meteor. Atmos. Phys.*, **59**, 65-82.

Lemon, L.R. and M. Umscheid, 2008: The Greensburg, Kansas tornadic storm: A storm of extremes. Preprints, 24th Conf. on Severe Local Storms, Savannah, GA, Amer. Meteor. Soc., 2.4.

Metcalf, J.I., 1995: Radar observations of changing orientations of hydrometeors in thunderstorms. *J. Appl. Meteor.*, **34**, 757-772.

Office of the Federal Coordinator of Meteorology: Federal Meteorological Handbook #11, Doppler Radar Meteorological Observations. This is the key publication for United States users that governs the use of the WSR-88D NEXRAD radar. Available in full at: <http://www.ofcm.gov/homepage/text/pubs.htm>.

_____, 2011: WSR-88D Tropical Cyclone Operations Plan.

Ryde, J. W., 1946: The attenuation of radar echoes produced at centimeter wavelengths by various meteorological phenomena. *Meteo. Factors in Radio Wave Propagation*, The Physical Society, London, 169-188. Though much of the work goes back to 1941, Ryde first predicted the possibility of weather radar forecasting.

Skolnik, M., 1980: *Radar Systems*. McGraw Hill, Second Edition.

Spratt, Scott M., David W. Sharp, Pat Welsh, Al Sandrik, Frank Alsheimer, Charlie Paxton, 1997: A WSR-88D Assessment of Tropical Cyclone Outer Rainband Tornadoes. *Wea. Forecasting*, **12**, 479–501.

Straka, Zrnic, Ryzhkov, 2000: Bulk hydrometeor classification and quantification using polarimetric radar data: Synthesis and Relations. *J. Appl. Meteor.*, **39**, 1341-1372.

Torres, S. M. and C. D. Curtis, 2007: Initial implementation of super-resolution data on the NEXRAD network. Preprints, 23rd Conf. on Interactive Information Processing Systems (IIPS) for Meteorology, Oceanography, and Hydrology, San Antonio, TX. The key paper that formally documented super-resolution WSR-88D data.

Torres, S. M. and C. D. Curtis, 2006: Design considerations for improved tornado detection using super-resolution data on the NEXRAD network. *Preprints, Third European Conf. on Radar Meteorology and Hydrology (ERAD)*, Barcelona, Spain, Copernicus. One of the very first papers to introduce the concept of super-resolution WSR-88D data.

Vivekanandan, Zrnic, Ellis, Oye, Ryzhkov, Straka, 1999: Cloud microphysical retrieval using S-band dual-polarization radar measurements. *Bull. Amer. Meteor. Soc.*, **80**, 381-388.

Whiton, Roger C. and Paul L. Smith, 1998: History of Operational Use of Weather Radar by U.S. Weather Services. Part I: The Pre-NEXRAD Era". *Wea. and Forecasting*, **13**, 219-243. One of the best summaries of the era of weather radar in the U.S. before the 1990s.

Zrnic, D. S., and A. Ryzhkov, 1999: Polarimetry for Weather Service Radars. *Bull. Amer. Met. Soc.*, 389-406

Index

Symbols

3-D 36

A

absorption 18
Air Force Cambridge Research Laboratories 6
amplitude 17
anomalous propagation 47
antenna
 WSR-88D 39
anticyclonic divergence 74
anticyclonic rotation 72
ash clouds 48
Asia 11
attenuation 18, 19
attenuation factor 22
Australia 12
azimuth 28

B

backscattering 19
base product. *See* product
bats 48
beam 26
beam blockage 28, 29
beam spreading 29
beamwidth 26
 and cone of silence 30
 effective 33
 physical 27, 33
bin 27, 32
birds 48
 illustration 129
bit level 36
bits
 and data resolution 36
bounded weak echo region. *See* BWER
bow echo 58
 and velocity data 76
Brazil 10
bright band 52
BUFR 33
BWER 56
 illustration 55

C

Canada 9
C-band 9, 16, 44
CC. *See* cross-correlation coefficient
CD. *See* contiguous Doppler
centroids 109
chaff 49
China 11
clutter mitigation decision 34
clutter reduction 34
CMD. *See* clutter mitigation decision
coalescence 51
color
 of velocity data 69
components 109
composite reflectivity
 and mosaics 37
 defined 103
cone
 shape of tilts 29
cone of silence 30
contiguous Doppler 65
contiguous surveillance 65
convective precipitation 119
convergence 73
Cornell University 6
couplet 71
CPS-9 3
CR. *See* composite reflectivity
cross-correlation coefficient
 defined 90
cross-polar power 88
cross section 34
CS. *See* contiguous surveillance
Cuba 10
cyclonic convergence 74
cyclonic rotation 72
CZ. *See* composite reflectivity

D

data
 radar 32
data windowing 33
dealiasing 65
decibel 17
depolarization 99
derived products 103
descending reflectivity core. *See* DRC
differential phase shift. *See* specific differential phase
differential propagation phase
 defined 92
differential reflectivity
 definition 88
diffraction
 and sidelobes 21
digitization
 of radar data 5
divergence 73
DMRL-C 11
doppler dilemma 65
Doppler radar
 dual 8
 history 6
downburst
 and velocity data 76
DRC 57
dryline 118
dual Doppler 8
dual pol. *See* polarimetric radar
ducting 49
DVIP. *See* Video Integrator and Processor

E

echo tops 105
electricity 1
electromagnetism 15
elevation 27, 28, 30
enhanced echo tops 105
equivalent reflectivity 22
 equation 22

E

ETVS. *See* tornadic vortex signature
Europe 10
evaporation 51

F

field
  electric 87
filler radars 9
frequency 16
front 118

G

gate 27
ground clutter 30, 34, 47

H

hail 97
  and diagnosis 120
Hail Detection Algorithm (HDA) 111
HCA. *See* hydrometeor classification algorithm
HDF5 33
hertz 16
Hertz, Heinrich 2
hook echo 56
human-computer interface (HCI) 40
hurricanes 125
hydrometeor classification algorithm 94, 124
Hz. *See* hertz

I

ice pellets 97
inbound
  defined 69
intensity 25
inverse square law 22
inversion 49
isodop 70
isopleth 70
isosurface 57

J

Japan 11

K

Kavouras 5
KDP. *See* specific differential phase

L

lambda 15
LDR. *See* linear depolarization ratio
Lemon, Les 53
Level I data 39
  composition 40
Level II data 32, 39
  composition 40
Level III data 32, 40
  composition 40
LEWP. *See* line echo wave pattern
linear depolarization ratio 93
listening period 22
localized features 38
Luke Range 49

M

Manually Digitized Radar 5
Marconi, Guglielmo 2
maximum estimated hail size 112
maximum unambiguous range 26
maximum unambiguous velocity 64
Maxwell, James Clerk 2
MCS 60
MDA. *See* mesocyclone detection algorithm
MDR. *See* Manually Digitized Radar
melting layer 94
melting layer detection algorithm 93
melting level 52, 98, 124
MESH 112
mesocyclone
  and diagnosis 121
  and velocity data 76
mesocyclone detection algorithm 110
mesoscale convective system 60
Mexico 9
Mie scattering. *See* scattering
MIT 3
MLDA. *See* melting layer detection algorithm
mosaic 37
mountains 30
MPAR. *See* multifunction phased array radar
multifunction phased array radar 8

N

narrowband 40
National Center for Atmospheric Research 6
National Severe Storms Laboratory 6
National Weather Service 8
NetCDF 33
Nevada Test and Training Range 49
New Zealand 12
NEXRAD. *See* WSR-88D
NOAA 32
Nyquist co-interval 65
Nyquist interval 65
Nyquist velocity 64

O

oblate 88
OPERA 11
Operational Program on the Exchange of Weather Radar 11
outbound
  defined 69
Outflow boundaries 118
oversampling 29

P

phase 18
phosphor 34
plane
  polarimetric 87
plan position indicator 34
  illustration 35
plan view 34
polarimetric radar
  differential diagnosis 97
  history 8
  introduction 87
polarity 8, 18
  illustration 19
POSH 112
power 17, 22

cross-polar 88
PPI. *See* plan position indicator
PPS. *See* precipitation processing subsystem
precipitation
  convective 119
  estimation 106
  stratiform 119
  winter 122
precipitation processing subsystem 107
PRF 25
principal user processor (WSR-88D). *See* PUP
probability of significant hail 112
product
  base 32
  derived 103
prolate 89
PRT 25
pulse 22
  length 25
  repetition time 25
pulse length 25
pulse-pair processing 18, 63
pulse repetition frequency 25
pulse repetition time 25
PUP 40

R

radar 1
  airborne 18
  filler 9
radar data acquisition unit (WSR-88D). *See* RDA
radar product generator unit (WSR-88D). *See* RPG
radar reflectivity factor 25
radial 27, 32
radial velocity 75
radio 1
radome
  interference with 50
  of WSR-88D 39
rain 97
range 27, 28
  maximum unambiguous 26
  range folding 26, 50
  and velocity data 65
range-height diagram 136

range-height indicator 34
  and cross sections 36
  illustration 35
RAREP 4
Rayleigh scattering. *See* scattering
RDA 39
rear inflow notch 59
reflectivity 22, 47
  and severe weather signatures 52
reflectivity factor 25
refraction 21
  subrefraction 49
  superrefraction 49
refreezing layer 95
resolution 28
  angular 28
  data 36
  radial 28
RHI. *See* range-height indicator
Rio Grande 48
RMAX 26, 50
Roshydromet 11
rotational velocity
  of couplet 71
RPG 40
Russia 11

S

San Antonio 48
S-band 9, 16
scan strategy 32
scattering 19
  and wavelength 21
  forward 19
  geometric 19
  Mie 19, 21
  optical 19
  Rayleigh 19, 21
second-trip echo 26, 50
segments 109
SELS. *See* Severe Local Storms Unit
Severe Local Storms Unit 4
shear
  of couplet 71
sidelobe 21, 30
  and WSR-88D 22
slant-45 87

smoothing 37
snow 97
South Africa 11
South Korea 11
specific differential phase 92
spectrum width
  defined 77
squall line 59
Storm Cell Identification and Tracking (SCIT) 109
storm relative velocity 75
storm total precipitation 108
stratiform precipitation 30, 119
subrefraction 49
sun strobe 50
supercell
  and hook echoes 56
superrefraction 49
super resolution
  and data windowing 33
  and smoothing 37
  definition 42

T

TBSS. *See* three-body scatter spike
TDA. *See* tornado detection algorithm
TDWR 43
Terminal Doppler Weather Radar. *See* TDWR
Tesla, Nikola 2
Thailand 11
three-body scatter spike 21
  illustration 20, 80
three-dimensional display. *See* 3-D
thunderstorm 52
  motion nomogram 137
  type 120
tilt 28, 30, 32
  of reflectivity core 53
tornadic vortex signature 111
tornado
  and diagnosis 121
  and velocity data 76
tornado debris 99
tornado detection algorithm 111
tornado warning 4
tropical cyclones 125
TVS. *See* tornadic vortex signature

U

Umscheid, Mike 57
unit control position (UCP) 40

V

VAD. *See* velocity azimuth display
VAD wind profile 80
VCP 32, 42
  table of 144
velocity
  Nyquist 64
  of light 15
  processing of 63
  radial 75
  sign 69
  storm relative 75
velocity azimuth display 80
vertically integrated liquid
  defined 104
Video Integrator and Processor 4
VIL. *See* vertically integrated liquid
VIP. *See* Video Integrator and Processor
VMAX 64
volcanoes 48
volume 32
volume coverage pattern. *See* VCP
volume scans 6
volumetric features 38
vortex hole 57
VWP. *See* VAD wind profile

W

warnings 4
wavelength 15
  and scattering 21
  millimeter 17
waves
  electromagnetic 15
WBRR. *See* Weather Bureau Radar Remote
weak echo region. *See* WER
Weather Bureau 4
Weather Bureau Radar Remote 5
WER 53
  illustration 54
wildfires 48
wind

  and diagnosis 121
windows, display 36
wind profile 79
wind turbine clutter. *See* WTC
winter weather 122
wireless 1
World Meteorological Organization 33
World War II 2
WSR-57 4, 7
WSR-74 7
WSR-88D
  antenna 39
  design 39
  development of 6
  polarimetric upgrade 42
  sidelobes 22
WTC 49
WWII. *See* World War II

Z

ZDR. *See* differential reflectivity
zero line 70

CPSIA information can be obtained
at www.ICGtesting.com
Printed in the USA
BVHW022236160119
538058BV00008B/59/P